Mars

by Robert Godwin

We acknowledge the financial support of the Government of Canada through the Book Publishing Industry Development Program for our publishing activities.

Published by Apogee Books, Box 62034, Burlington,
Ontario, Canada, L7R 4K2, http://www.apogeebooks.com
Tel: (905) 637-5737

Printed and bound in Canada

Mars - Pocket Space Guide by Robert Godwin
Apogee Books Pocket Space Guide #2

ISBN-10: 1-894959-26-4
ISBN-13: 978-1-894959-26-1

Mars Facts

General

One of five planets known to the ancients; Mars was the Roman god of war and agriculture

Yellowish brown to reddish color; occasionally the third brightest object in the night sky after the Moon and Venus, currently believed to be about 4.6 billion years old

Physical Characteristics

Average diameter 6,780 kilometers (4,212 miles); about half the size of Earth, but twice the size of Earth's Moon

Same land area as Earth, reminiscent of a rocky desert

Mass 1/10th of Earth's; gravity only 38 percent as strong as Earth's

Density 3.9 times greater than water (compared to Earth's 5.5 x greater than water)

No planet-wide magnetic field detected; only localized ancient remnant fields in various regions

Orbit

Fourth planet from the Sun, the next beyond Earth

About 1.5 times farther from the Sun than Earth is

Orbit elliptical; distance from Sun varies from a minimum of 206.7 million kilometers (128.4 millions miles) to a maximum of 249.2 million kilometers (154.8 million miles); average distance from the Sun 227.7 million kilometers (141.5 million miles)

The focus of Mars' orbit is about 13 million miles from the Sun

Revolves around Sun once every 687 Earth days

Rotation period (length of day) 24 hours, 39 min, 35 sec (1.027 Earth days)

Poles tilted 25 degrees, creating seasons similar to Earth's

Environment

Atmosphere composed chiefly of carbon dioxide (95.3%), nitrogen (2.7%) and argon (1.6%)

Surface atmospheric pressure less than 1/100th that of Earth's average

Surface winds up to 80 miles per hour (40 meters per second)

Local, regional and global dust storms; also whirlwinds called dust devils

Surface temperature averages -53° C (-64° F); varies from -128° C (-199° F) during polar night to 27° C (80° F) at equator during midday at closest point in orbit to Sun

Features

Highest point is Olympus Mons, a huge shield volcano about 26 kilometers (16 miles) high and 600 kilometers (370 miles) across; has about the same area as Arizona

Canyon system ofValles Marineris is largest and deepest known in solar system; extends more than 4,000 kilometers (2,500 miles) and has 5 to 10 kilometers (3 to 6 miles) relief from floors to tops of surrounding plateaus

"Canals" observed by Giovanni Schiaparelli and Percival Lowell about 100 years ago were a visual illusion in which dark areas appeared connected by lines. The Mariner 9 and Viking missions of the 1970s, however, established that Mars has channels possibly cut by ancient rivers

Evidence of ancient salty seas discovered in 2004 by on-site Rovers

Moons

Two irregularly shaped moons, each only a few kilometers wide

Larger moon named Phobos ("fear"); smaller is Deimos ("terror"), named for attributes personified in Greek mythology as sons of the god of war

Mars - Pocket Guide

MARS

In early Roman folklore Mars was the god of fertility and vegetation. Our ancestors looked to Mars to bring healthy crops to their fields. Later Mars would become the god of death and war. Over nearly four centuries of observation the planet Mars would undergo a similar transformation, from what was believed to be a lush fertile oasis to a place totally hostile to life as we know it. This transformation took place because of the dedicated and tireless efforts of generations of astronomers and scientists.

An important part of this journey of discovery began almost three hundred years ago when an Italian astronomer named Jean Dominique Maraldi turned a primitive telescope toward the red planet. Maraldi had been studying our closest planetary neighbor since 1672 and he had confirmed that Mars rotates on its axis once in just over 24 hours. Now he perceived that there were white patches at the north and south Martian poles. He noticed that they seemed to shrink and grow in accordance with Mars' passage around the Sun. Due to a remarkable coincidence it seemed that not only was a Martian day about as long as an Earth day, but the red planet was not spinning perfectly vertically as it journeyed around the sun. It was tilted—just like the Earth—and therefore it had seasons.

Later that century an accomplished astronomer named William Herschel took his turn studying Mars and confirmed the waxing and waning of the white spots. He concluded that what he was seeing were in fact ice-caps.

William Herschel

This was an enormously important discovery, particularly since at that time, 1769-1783, no one had even explored the Earth's polar regions. They were inaccessible and the one invention which might conceivably allow polar exploration, the balloon, would not be invented by the Montgolfier brothers until late in 1783. Remarkably, inquiring human eyes were exploring the Martian poles in better detail than they had ever explored their terrestrial equivalent. Herschel further refined the length of Mars' day, concluding that it was 24 hours, 39 minutes and about 20 seconds.

A little more than a decade later, in 1796, a renowned French scientist called Pierre Simon de Laplace published a book called *Exposition du systeme du monde* (Exposition of the system of the world). In this book Laplace combined his considerable mathematical skills with Newton's gravitational theories in an attempt to explain how the solar system was formed. Laplace reasoned that an immense cloud of gas had gradually been drawn together by gravity to create an enormous whirlpool. As the cloud grew, the rotation of the whirlpool spun faster. The particles of gas and dust clustered together in the middle and formed the Sun. As the whirlpool sucked material into its center, tufts of dust and gas were flung off in great doughnut shaped rings which eventually clumped into planets.

Pierre Simon de Laplace

If Laplace was correct, then the outer planets must have formed first. Therefore, Mars was older than the Earth. It was a brilliant theory, and much of it is still considered correct. The best evidence available today make Mars and the Earth about the same age. But if Laplace was right and

Mars was older, then by extension if there was life on Mars it must also be older and therefore more advanced.

As telescopes became more powerful and the optics became more refined, the images of Mars became clearer. Science continued to undergo an unprecedented revolution during the 19th century and the revelations from the astronomers seemed to be increasingly spectacular. The universe seemed to expand inexorably as telescopes became more powerful—an apparent expansion that continues even to this very day. Inevitably this spurred a debate between cosmologists and theologians about man's place in the cosmos. Were we alone?

Every day seemed to bring more startling discoveries and the universe was soon considered to be no longer just the imagined reverie of a benevolent supreme being but a tangible, infinite, physical form. Something that humanity might actually be able to explore. A place where we might encounter strange neighbors—other citizens of the cosmos. Whatever your personal beliefs, there was no longer any reason to think we might not have company in the vast and seemingly infinite depths of space. Many brilliant and articulate philosophers and scientists debated this important issue. Those who thought we were alone made persuasive arguments in the face of the staggering onslaught of statistics. It is the same argument which rages today; although the voices raised on both sides back then seemed to be more willing to listen to each other than they are today.

Are we alone? It is the siren-song which has driven human space explorers to risk their lives in search of an answer. Many robotic craft have been dashed against the Martian rocks and yet some few have managed to successfully land on Martia Firma, their controllers employing all the cunning of Ulysses to

Pietro Angelo Secchi

avoid the tangle of dangers that lie between here and our sister planet.

In 1858 an astronomer working at the Vatican observatory named Father Pietro Angelo Secchi took it upon himself to create his own drawings of Mars. The red planet was now nearing a close approach to Earth and the powerful Vatican telescope was capable of resolving detail previously invisible to most astronomers. Secchi thought he saw a series of straight lines on the Martian surface so he made an innocuous notation in his notes. His sketches and article were published in 1859 in which he referred several times to "Canale Atlantico" or "Canale ceruleo". His regrettable choice of words would not have an impact for another eight years.

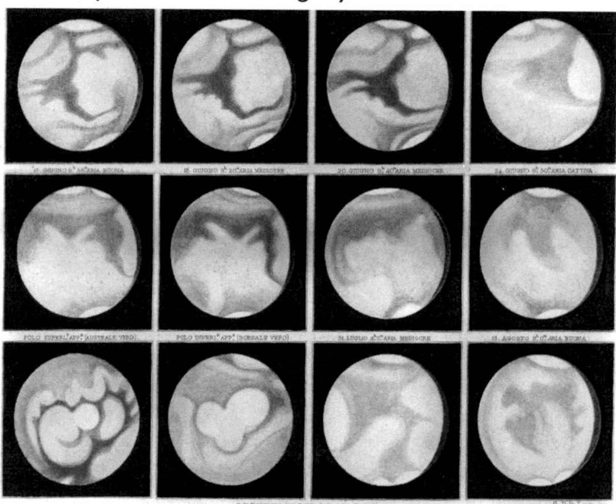
Father Pietro Secchi's drawings of Mars from 1858

In 1871 the possibility of Martian life was briefly enhanced when a scientist called William Huggins turned the recently devised spectroscope on Mars and detected the presence of water vapor. Another brilliant astronomer called Christian Huygens had recognized as early as 1695 that life needed some kind of solvent to exist. On Earth the solvent is water and, according to Huggins, there was water on Mars.

William Huggins

The discovery of Martian water spurred the academic debate to new heights. Less than three months after Huggins' report a respected astronomer in England named Richard Proctor tackled the issue head-on in the mainstream press. He observed this about the Martian polar ice-caps, *"Here, then we have a feature which we should certainly expect to find if the polar spots are really snow-caps; for the existence of water in quantities sufficient to account for snow regions covering many thousand square miles of the surface of Mars would undoubtedly lead us to infer the existence of oceans."*

He then addressed the concept of the reduced Martian gravity, *"The force of gravity is so small at the surface of Mars that a mass which on the Earth weighs a pound, would weigh on Mars but about six and a quarter ounces...A being shaped as men are, but fourteen feet high, would be as active as a man six feet high, and many times more powerful. On such a*

Richard Proctor

scale then, might the Martial navvies be built. But that is not all. The soil in which they work would weigh very much less, mass for mass, than that in which our terrestrial spadesmen labour. So that, between the far greater powers of Martial beings, and the far greater lightness of the materials they would have to deal with in constructing roads, canals, bridges, or the like, we may very reasonably conclude that the progress of such labours must be very much more rapid, and their scale very much more important, than in the case of our own Earth."

This clearly illustrates that the consensus of thought was beginning to lean toward the possibility of huge artificial constructs on Mars. A seed of extreme possibility had been sown.

Two years later Proctor was gradually being convinced that Martian life was perhaps less likely than he at first believed. Now he realized that if Mars had such an abundance of water in its atmosphere there would be evidence of an equally abundant precipitation of snow in the winter. Since Mars is not white he concluded that the Martian atmosphere must be much more tenuous than previously suspected. He noted that the atmospheric pressure may be equivalent to several miles above the earth's surface and therefore inadequate to support animal life. He did acknowledge that his French counterpart, Camille Flammarion, had noted the abundance of bacteria on Earth which reside even at altitudes abhorrent to more complex life. One of his more startling predictions reads, *"There cannot, then be anything like the accumulation of snow which gathers in regions above our snow-line; but instead of this there must exist over the surface of Mars except near the poles a thin coating of snow, or rather there will be ordinarily a mere coating of hoar frost."*

This indeed was proven to be the case when the Viking landers spent their first winter on Mars in 1976 and took pictures of a tenuous layer of frost condensing out of the thin Martian air. In one final droll footnote Proctor stated that life on Mars would be inconvenient since the low air pressure would make it impossible to boil water to make a decent cup of tea or a well-boiled potato. *"It does not make matters more pleasant that the tea-plant and the potato are impossible of themselves on Mars, and that the possibility of boiling them may be regarded as a secondary consideration...Our hardiest forms of vegetable life would not live a single hour if they could be transplanted to Mars."*

Two more years would pass, during which time more information poured in from the world's observatories and then in June 1875 Proctor once more revisited the pages of the *Cornhill* magazine with another enlightened article called *"Life, Past and Future in Other Worlds."* He explained his change of heart regarding his position on Martian life and made a few more interesting observations that still prosper today. He said that Mars must have once been like the Earth, warmer, wetter, volcanically active and generally more accommodating to life. He then suggested that, *"Mars was a world like our own, filled with various forms of life. Doubtless these forms changed as the conditions changed around them...finally perishing as cold and death seized the planet."* These were strong words coming from such an eminent astronomer.

All eyes continued to watch Mars as it gradually drew closer to the Earth and the best telescopes in the world were trained on the approaching planet. Astronomers competed to make new discoveries. Indeed, a contest had broken out over who had naming rights over each new discovery. Richard Proctor in England used his own impenetrable logic to name no less than four features after the Reverend William Dawes, a fellow astronomer, while Camille Flammarion in France applied his own names. An American astronomer by the name of Asaph

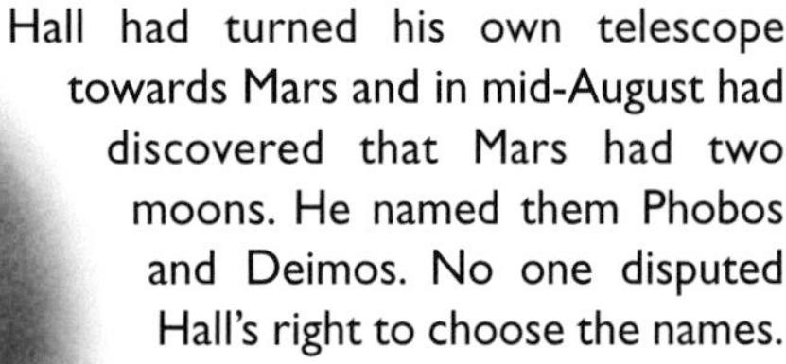

Asaph Hall

Hall had turned his own telescope towards Mars and in mid-August had discovered that Mars had two moons. He named them Phobos and Deimos. No one disputed Hall's right to choose the names.

Ten years of argument after Proctor had first named the features on Mars, a color-blind Italian with a sharp eye and the patience of Job resolved the debate about Martian nomenclature. Giovanni Virginio Schiaparelli was born in 1835, in the town of Savigliano in northwestern Italy. At an early age Schiaparelli showed an enthusiasm for sky-watching thanks in some part to the encouragement of a local priest. By the end of 1874 Schiaparelli had drawn so much attention to himself that he was given a new, more powerful telescope to continue his observations. The timing proved quite fortuitous as he then turned his attention to the planets. Mars was closer than usual and this provided an unprecedented opportunity for new discoveries.

Giovanni Schiaparelli

The Earth's orbit around the sun is not circular, it is very slightly elliptical. This means that at its closest approach to the sun it is 3 million miles closer than at its furthest point; this discrepancy amounts to just over 3%. Mars' orbit, however, has a much more pronounced elliptical aspect. The difference between its closest and furthest approaches to the sun is over 8%. The net effect of this is that Mars and the Earth come into close alignment at irregular intervals. The nearest estimate is that the two come into similar relative positions about every 284 years. However, approximately every two years or so, the two planets pass one another on the same side of the sun, this is called an "Opposition" or "Conjunction". Naturally the best time to turn a telescope on Mars would be at the time of just such an Opposition.

At about the same time as Hall's startling discovery, back in Europe, Schiaparelli became keen to test his new telescope's ability to discern the features on Mars' surface. It was not his initial intention to spend a great deal of time on this but nonetheless he wanted to compare notes with those who had preceded him. At first glance he was not very encouraged by what he saw, *"I must confess that, on comparing the aspects of the planet with the maps that had been most recently published, my first attempt did not seem very encouraging."*

By September 12th Schiaparelli decided to compile his own map of the Martian surface. His meticulous nature and talents soon swept him to the forefront of Mars' observers when his new map was published in 1878. Probably one of the smartest things he had done was to rename all of the major features using Latin names. He had done this because most of his drawings didn't coincide with those on existing maps and so he found himself compelled to adopt a new nomenclature. Purely by accident his unassuming method had put an end to the long running naming dispute.

The success of Schiaparelli's map of Mars would lead him to spend much of his time continuing its refinement. Some of his

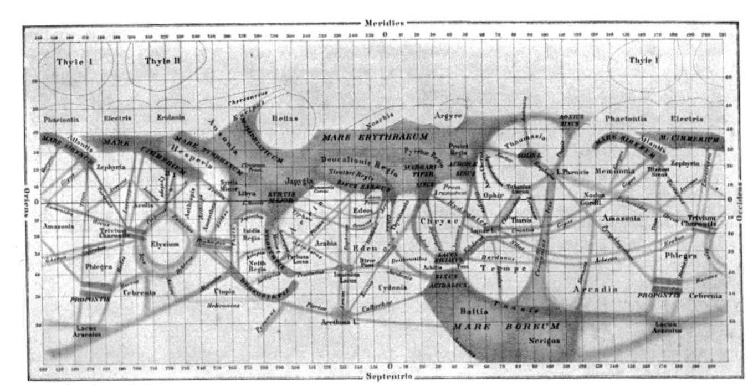

One of Schiaparelli's maps of Mars

later maps were drawn in markedly different ways but the one defining thing that would remain was his assertion that Mars was covered in a fine latticework of grooves and straight lines. The Italian word for channels is "canale." Schiaparelli would see many more of these lines than his predecessor Secchi and he would use the same word to describe them.

Now, not for a single moment did Schiaparelli suggest that this latticework was a fabrication made by Martians. His best guess was that the channels were natural formations that appeared during the Spring thaw on Mars and were caused by the melting of polar ice. However, his unfortunate decision to use the same word as Secchi would lead to a dispute that would completely overshadow not only the debate about Martian names but also Schiaparelli's great accomplishments.

When you translate "canale" into French, Spanish or German you get something equivalent to "canals". Of course the English is virtually identical. The implication of a canal is that it has been contrived by someone to move water (and perhaps vehicles) from one place to another. Unfortunately for scientists, and particularly for Schiaparelli, the world was still a pretty ignorant place, especially in the understanding of science, and people were itching for something to spark their imagination. Schiaparelli's canals were just what they needed. The little-green-man had been born…

The general populace became completely enamored with the Martians. What were they like? Why did they build these giant waterways? Could we communicate with them? Were they friendly? If nothing else Secchi and Schiaparelli's regrettable choice of words ultimately caused the whole world to look toward the heavens like never before

In the summer of 1877 with Mars a bright object in the evening sky, discussion about life on the red planet was at an all-time high. Camille Flammarion the noted French astronomer suggested that the waxing and waning of the surface features of Mars might be explained by vegetation. He also proposed that the vegetation might well be red rather than green. As ever, Richard Proctor chimed in, *"Unless we adopt the theory that not only is the vegetation of Mars rubescent, but that all the principal glories of the Martian forests are ever-reds, and the Martian fields covered with herbage of unchanging ruddiness, we must accept the conclusion that the land surface is an arid desert. This evidence alone is almost strong enough to assure us that none but the lowest forms of life, animal and vegetable, exist on Mars at present. The evidence against the fitness of Mars to support the higher forms of life seems overwhelmingly strong."*

Camille Flammarion

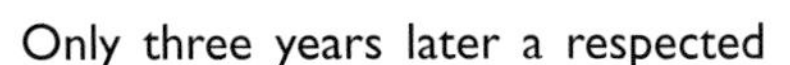

Only three years later a respected

social critic by the name of Percy Greg wrote one of the first novels to introduce an encounter with the Martians. The book was called *Across the Zodiac* and it is one of the first to fully explore the social morals of an alien society. Greg is the first to use the word *astronaut*, although it is the name of the space vehicle rather than the passenger, and he discusses how his hero has almost superhuman capabilities while he is on Mars, by virtue of the reduced gravity. He takes full advantage of Schiaparelli's canals even to the point of a romantic voyage across Mars with his hero and love-interest indulging in some sightseeing. Greg is also the first to suggest the devastating effect that cross-planetary infections might cause when he has one of his Martian characters succumb to a terrestrial disease.

It is perhaps a justifiable irony that Schiaparelli's name is not the one which most people associate with Martian canals. That dubious privilege falls to a graduate of Harvard University by the name of Percival Lowell. He was born in 1855, in New England, and was particularly drawn to Schiaparelli's assertions of Martian symmetries. Seventeen years after the famous Schiaparelli maps appeared he established an observatory in the high desert above Flagstaff Arizona. It was a perfect location for good "seeing" due to the still desert air and its 7000 foot altitude. Within less than a year Lowell was ready to publish his observations of Mars in the noted publication *Atlantic Monthly*.

Percival Lowell

Beginning in May of 1895 and continuing through August, Lowell submitted the basic text which would form his first book simply titled

Mars. Atlantic Monthly was already well established as one of America's better magazines, running everything from political commentary to fiction and science. It provided an instantaneous and wide forum for Lowell's contentious observations. Lowell was convinced that the Martian climate was in fact much milder than previously supposed. He attributed it to a modicum of water vapor in the Martian atmosphere, "*...for aqueous vapor is quite specific as a planetary comforter, being the very best of blankets. It acts, indeed, like the glass of a conservatory, letting the light rays in, and opposing the passage of the heat rays out.*"

After some theoretical analysis of the Martian climate Lowell supposed that the atmospheric pressure was insufficient to support clouds in a form like those on Earth. He noted that several of his contemporaries were observing ephemeral bright patches, but he could not attribute them to clouds because of his belief that the Martian mountains were quite small and could not disturb the air sufficiently to create clouds. Of course what Lowell could not have known is the absurd altitude of Mars' volcanoes. One such structure, affectionately known as Mount Olympus, is in fact fifteen miles high.

Lowell was happy to concede that Mars' atmosphere was likely extremely thin, and of course in this assessment he was correct. He did not, however, preclude life on this basis. "*That beings constituted physically as we are would find it a most uncomfortable habitat is pretty certain.*"

Continuing his observations at Flagstaff with his new telescope, Lowell monitored Mars for much of the summer of 1894. What he saw must have seemed quite extraordinary. Sights previously not seen or noted by human eyes. As the

Martian summer proceeded, the southern polar cap was seen to melt.

It is important to draw a distinction between melting and sublimating. What in fact happens on Mars is that the large amounts of water vapor and frozen carbon dioxide locked up in the titanic southern ice-cap sublimate out into the atmosphere. It is generally accepted today that there exists very little possibility of liquid water on the surface of Mars. The air is just too thin and as the seasons change the ice simply vaporizes. What Lowell was sure he saw, and many of his colleagues agreed with him, was a Martian polar sea. Professor William Pickering, who also worked at Flagstaff, turned a polariscope on Mars and seeing that the returning light was polarized (i.e. organized in such a way that it behaves differently to regular light) he concluded that it must indeed be liquid water he was seeing. (Sunlight bouncing off water is polarized.)

Lowell began a concerted effort to study the spread of the Martian seas, working from May until November he recorded a host of changes in the color of the surface. At this point there was little or no reference to the linear structures which Schiaparelli had documented. Most of Lowell's comments are about color changes, specifically from orange-ochre to blue-green. He noticed that the color changes seemed to coincide with the seasonal tilt of the planet and thus with its seasonal weather. When the blue-green regions suddenly vanished Lowell was forced to conclude that if he had been seeing lakes or small seas then he should be able to see where the water had gone. Since nothing leaped out, such as an enlarged polar cap, he had to rethink.

"There is thus reason to believe that the blue-green regions of Mars are not water, but generally at least, areas of vegetation; from which

it follows that Mars is very badly off for water, and that the planet is dependent on the melting of its polar snows for practically its whole supply."

Lowell defended the existence of artificial waterways with increasing fervor, *"Although skepticism as to the existence of the so-called canals seems now pretty well dispelled, disbelief still makes a desperate stand against their peculiar appearance...for that they are both straight and double, as described, is certain—a statement I make after having seen them instead of before doing so, as is the case with the gifted objectors."*

Schiaparelli himself supposed the canals to be natural formations but when asked if he thought they might be artificial he stated, *"I should carefully refrain from combating this supposition, which involves no impossibility."* Spoken like a true scientist.

As has been demonstrated, the impact of the close opposition of Mars created a cascade of interest in the possibility of alien, and specifically Martian life. Scientists and fiction writers alike were bursting at the seams with new ideas. The combination of Schiaparelli's canals and Hall's dual moons were ripe material for romantic fiction of an entirely new sort. While Percy Greg used his considerable skills to reprimand the authorities in England, Camille Flammarion, Richard Proctor and Giovanni Schiaparelli took advantage of this new interest to educate the general population about Mars. Magazine editors in Europe and America soon sought after all three.

In April of 1897 a highly regarded novelist from south London unleashed his latest fiction on an unsuspecting British and American public. His name was Herbert George Wells and the story was called *The War of the Worlds*. Wells created a torrid, war-torn landscape in which Britain is overrun by enormous

Herbert George Wells

leviathans on stilts wielding a creation which seems to have been entirely original, the heat ray. Wells' hero staggers from one disaster to the next, always only just one step ahead of the invisible and god-like aggressors. It is a chilling and sparsely written story by today's standards. There are few characters, little dialogue and almost no respite from the horror. The prose almost drips with paranoia and fear, and the reader is dragged along by Wells economical language into a world unlike any before it—a familiar one, but one that is controlled by creatures from another world. Wells himself admitted that the idea came about when he was taking a leisurely walk in the Surrey countryside with his brother Frank. He remembered chatting

FIGHTERS FROM MARS,

OR

The Terrible War of the Worlds as It Was Waged in and Near Boston in the Year 1900.

Startling.
Thrilling.
Extraordinary.

Absolutely Unlike Any Story Ever Before Written.

Destruction of the Boston City Hall by the Strange Visitors From Mars.

Right of Exclusive Reproduction in This Part of the Country Secured by Cable From the Author, Mr. H. G. Wells.

Begins in the Sunday Post Next Sunday, Jan. 9.

An 1898 Boston newspaper advertisement for *War of the Worlds* about the intervention of Europeans in Tasmania and the subsequent reckless destruction of the native populations. Frank surmised, *"Suppose some beings from another planet were to drop out of the sky suddenly, and began laying about them here!"* Wells enjoyed the "realizing of disregarded possibilities" and so he reckoned, *"There may be life on Mars—that planet could support life, and that life would have to obey certain conditions; intelligence may have gone farther there than on this planet."*

Garrett Putnam Serviss

Wells' story would be serialized, first in popular magazines and then in the newspapers of William Randolph Hearst where it would appear every day during the months of January and February 1898. It was promptly followed up by an unofficial sequel commissioned by Hearst and written by his science editor, the noted astronomer Garrett Serviss. The sequel was called *Edison's Conquest of Mars* and the two stories would ignite the imagination of a young boy growing up in the conservative ambience of New England. His name was Robert Goddard and the following year, in October

One of the earliest drawings of a spacesuit from *Edison's Conquest of Mars*

1899, he had a daydream while climbing the cherry tree in his backyard. He dreamed of flying to Mars in a spaceship of his own design.

Over the next two decades the study of Mars continued unabated. In 1906 Goddard proposed the use of nuclear propulsion as a means to send a craft into space but his article was refused by *Popular Astronomy* magazine with the following comment from the editor, "*Let me say briefly about the paper 'On the Possibility of Navigating Interplanetary Space,' that the possibility by your showing is so remote is it worthwhile to publish it? The speculation about it is interesting, but the impossibility of ever doing it is so certain that it is not practically useful. You have written well and clearly, but not helpfully to science as I see it.*"

The following year Alfred Russell Wallace took Lowell to task. Wallace thought Lowell was dead wrong and should have known better. He wrote, "*The water vapor of our atmosphere is derived from the enormous area of our seas, oceans, lakes and rivers, as well as from the evaporation from heated lands and tropical forests of much of the moisture produced by frequent and abundant rains. All these sources of supply are admittedly absent from Mars, which has no permanent bodies of water, no rain, and tropical regions which are almost entirely desert. Many writers have therefore doubted the existence of water in any form upon this planet, supposing that the snow-caps are not formed of frozen water but of carbon-dioxide, or some other heavy gas, in a frozen state.*" Assuredly this is the

Alfred Russell Wallace

most accurate description of Mars that had been written up to that point.

Robert Goddard

So it would seem that Wallace and others had shown that Mars was dead, and Goddard was being told his theories of propulsion for getting there were useless. The debate on Martian life should have finally been concluded. There was nothing there of interest and, even if there had been, no way to get there. For centuries Mars had been recognized as another world, people had dreamed of meeting Martians, and most recently people like H.G. Wells had supposed that the Martians might covet our world. It would be reasonable to assume that intellectuals like Wallace should have won this debate. But it was not over yet. In fact it wouldn't be over conclusively for another 57 years.

Many people really wanted to believe that we were not alone and that alien life was close-by—just next door. The Martian invader was still only an adolescent in the eyes of the general public and their misconceptions were still awaiting full exploitation, meanwhile Percival Lowell was still working the lecture circuit to record crowds.

The popularity of cheap pulp fiction was swelling to unprecedented proportions when a down-and-out pencil sharpener salesman decided he could write a story at least as good as some of the hackneyed bilge that appeared on the newsstands. He was 36 years old and he was obliged to place advertisements for his products in the cheap magazines of the day.

When he grew tired of checking the copy he would pore over the main content of the magazines. He soon realized that people were being paid substantial amounts of money to provide fantastic stories. His name was Edgar Rice Burroughs and he would take the world of pulp fiction and turn it completely on its ear. Not since H.G. Wells would a writer have such an impact, and his success would inevitably draw the public attention back to the red planet, Mars.

In February of 1912 Frank Munsey's *All Story Magazine* published a story called *Under The Moons of Mars*. It was the opening volley from Burroughs which would inaugurate his ascent to world-stardom. He had submitted the story under a pseudonym. When a price had finally been agreed for publication the story was to appear under the name *Normal Bean*. Apparently this was Burroughs' version of a disclaimer, a reassurance that he wasn't crazy, even though the story was so outrageously flamboyant. This ruse backfired when an over-zealous copy-checker changed it and the story appeared as written by ***Norman** Bean*. It would not be until the story was concluded in the July issue that the editor at *All-Story* would decide to reveal the identity of his newly discovered prodigy.

Edgar Rice Burroughs

The hero of the story was John Carter, a soldier in the Confederate American army. He is returning from a confrontation with a tribe of native Americans when he seeks refuge in a cave. Carter falls asleep in the cave and when he wakes up, he is on Mars. His method of conveyance is a kind of spiritual transference, a device used earlier by

THE ALL-STORY

VOL. XXII FEBRUARY, 1912. No. 2

Under the Moons of Mars
by
Norman Bean

several other writers. Burroughs barely concerned himself at all with explaining Carter's remarkable voyage, he just dove straight into the adventure.

Burroughs' Mars is populated by two distinct species, a nomadic warrior race (these are the giant green-skinned Martians) and a more technologically advanced race of red-skinned humanoids. They call their planet Barsoom and it is dry and in dire need of resources. Life on Barsoom is difficult.

Burroughs really makes little progress when it comes to accurately portraying Mars. He subscribes to the current scientific position that all the mountains must be small (less than four thousand feet) and he concocts a suitably daft explanation for the presence of a breathable Martian atmosphere. Techno-babble enthusiasts would definitely approve. His exposition of the operation of the Martian flying machines is equally impenetrable. It is, however, absolutely undeniable the impact that these stories made.

Meanwhile, less than a month after the appearance of John Carter a respected German Baron had a son, he called him

Hugo Gernsback

Wernher and his life-adventure would eclipse even that of the fictional John Carter.

Over the next few years the world looked inward as the worst war in human history was waged between the great colonial powers. Few people had time to look at Mars although a young entrepreneur called Hugo Gernsback would expend quite a few pages in his many magazines discussing the likelihood of Martian life. America would not join the European conflict until 1917 and in the meantime Gernsback's science magazines (called *Electrical Experimenter* and later *Science & Invention)* epitomized the atmosphere of enthusiasm for technological solutions prevailing in the United States. Woven in amongst articles on the latest weaponry to fight the war in Europe were suggestions for communicating with the Martians using everything from very high-powered searchlights to planting acres of crops in geometric patterns to attract the eye of any potential Martian observers. It was the crop-circle phenomenon, but in reverse.

Electrical Experimenter

Vol. VI. Whole No. 72 APRIL, 1919 No. 12

EDITORIAL

Interplanetary Messages

Once the war in Europe concluded in 1918, an era of brash enthusiasm ensued which lasted until the great depression at the end of the 1920's. The era of flight had arrived with a vengeance; since the war had forced governments to improve and perfect the flying machine. Following the war the young pilots who had undertaken so many hazardous missions returned to America and became famous as barnstormers. In 1929 one of these young men became an overnight sensation when he flew single-handedly over the Atlantic ocean. His name was Charles Lindbergh and the fame and fortune which he garnered for his magnificent feat would place him on a course which would ultimately lead him to Robert Goddard's doorstep. Lindbergh could see that Goddard was a genius and the young pilot was enthralled with the primitive rocket engines the inventor had built in his workshop in Roswell New Mexico. Lindbergh used his considerable political influence to persuade the Guggenheim foundation to fund the pioneer's experiments. Goddard had made enormously important breakthroughs in the 1920's, with some support from the Smithsonian institution, leading to his first successful launch of a liquid fueled rocket in 1926. Now with the not in-considerable financial support of Guggenheim he worked tirelessly in his pursuit of a means to explore the solar system. Just before Lindbergh's famous flight Goddard had re-read *The War of the Worlds* and *Edison's Conquest of Mars*. The young boy was now a man, but the red planet was still his ultimate goal.

On Mars
GERNSBACK

A long discussion about Martians appeared in August 1924 in Gernsback's *Science & Invention*

Goddard's ideas for rocket propulsion gradually seeped out into the consciousness of the scientific community and would not escape the attention of that enthusiastic young son of Baron von Braun in Germany. By 1932 Wernher von Braun had graduated from amateur experimentation to working full-time for the German army on rocket research. While Goddard worked almost alone, von Braun was surrounded with helpers and a sizeable budget. When the depression impacted the world's economy the money had gradually dried up from the Guggenheim's and Goddard was always left wondering when, or if, his next funds would arrive.

In Germany, Adolf Hitler had risen swiftly to power and he had set his eyes on dominating Europe, rocketry was to become a significant part of Hitler's plan to terrorize his neighbors. By the fall of 1938 the world was already becoming aware of Hitler's grim reality.

Into the drab and cheerless environment of the late 1930's came another young genius with an idea for a harmless prank. His name was Orson Welles and on Halloween eve of 1938 his relatively unknown Mercury theatre group unleashed the power of radio in a way that has never since been surpassed. Welles decided to create a dramatization of *The War of the Worlds* but he restaged the story in rural New Jersey. Welles used the full capability of radio with sound effects and a team of excellent voice actors to make the entire thing sound as though it were really happening. The effects of his prank were nothing short of

Orson Welles

astounding. By all accounts thousands of people who heard the broadcast actually believed it was a real news report. Many of these people didn't stay at their radios long enough to hear the disclaimers that Welles announced late in the hour. They ran panicked into the streets and fled from their homes—away from the onslaught of the mythical invading Martians.

The next day the Mercury Theatre players were instant celebrities and a visibly shaken Welles appeared before the press to apologize for any distress he may have caused. What is the most revealing part of this story is that the people of 1938 actually believed that a superior Martian civilization was not only real, but even had the wherewithal and the means to rampage across New Jersey, defeating the best that mankind could throw at them. People alive today lived in a world that to a great degree believed in technologically advanced Martians.

Within four years of Welles' broadcast the first missiles were on their way into space but their target was not Mars, it was London and Holland. Wernher von Braun's rocket team had created the world's first ballistic missile in late 1942. Sadly, even though the German scientists stated goal was space exploration the regime for which they worked had terrestrial targets in mind. For the next two years 3500 V-2 missiles rained death and destruction on the enemies of Hitler's Third Reich.

At the end of World War II the captured rocket science team, and its leader Wernher von Braun, were shipped off to the desert of New Mexico where they were obliged to demonstrate everything they knew about rocketry to their American counterparts. Robert Goddard, who was very ill and had little time left to live, is known to have been invited to a military

Theodore von Karman

depot after the conclusion of hostilities in Europe to see a captured V2. He is said to have recognized many of his own inventions and was only impressed by the V-2 in that someone had been given enough money to actually build it. If he had received similar support back when people accused him of being crazy, the war and the course of history would likely have been substantially changed.

An American aerodynamicist of Hungarian descent was invited to attend a launch of one of von Braun's captured V-2s. His name was Theodore von Karman and he had helped to establish a missile research department at a California university on behalf of the US Army ordnance department. It would later become known as the Jet Propulsion Laboratory. Another immigrant, from New Zealand, named William Pickering (not the same William Pickering that worked with Percival Lowell) joined von Karman and by 1954 he was the director of the facility. It is unlikely that Pickering or von Karman could have known at the end of the 1940's what an impact JPL would have on planetary exploration. Their main goals at the time were building rockets and jet engines and also perfecting the new science of telemetry.

Wernher von Braun

While von Braun struggled in confinement at White Sands, New Mexico and Pickering and von Karman were building JPL in California, some of the German rocket scientists had been recruited by other countries, notably Britain, France and especially the Soviet Union. The Russians quickly combined the talents of their own rocket pioneers with the considerable wealth of expertise captured from Germany and began a missile program which was unrivalled for two decades after the conclusion of World War II.

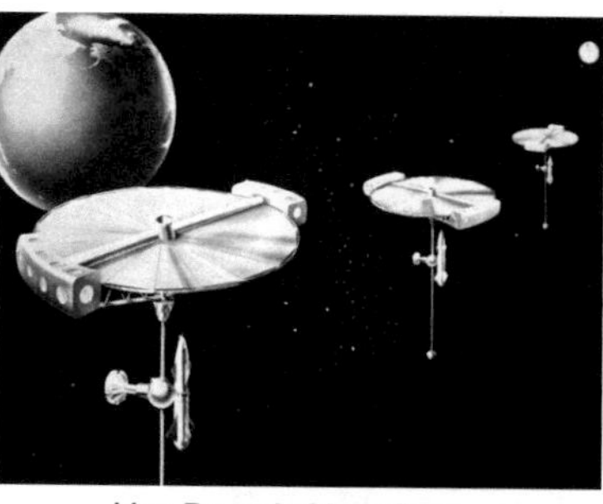

Von Braun's Mars fleet

In the mid to late 1950's Wernher von Braun had persuaded Walt Disney to join in his campaign to convince the American public that their future was in space. Through a series of articles in Collier's magazine, some follow-up books published by Viking Press and then finally a series of brilliantly rendered Disney animations, von Braun and Disney made a convincing argument for an exciting future for humanity in space. In the Disney movie Mars & Beyond, von Braun showed off his plans for a fleet of huge ion-propelled manned Mars spacecraft.

By 1957 the Soviet's chief designer, Sergei Korolev, had successfully built the biggest missile in the world. In October of that year he would use

Sergei Korolev

that rocket to herald the birth of the space age when he placed a small metal sphere called *Sputnik* into orbit around the earth.

By May of 1961 two humans had finally flown in space and bold new goals had been outlined by President Kennedy. During his speech to Congress in May of that year he not only set a challenge to send humans to the moon before the end of the decade, he also intimated that bigger aspirations were in his mind. After outlining his lunar goal he said, *"Secondly, an additional 23 million dollars, together with 7 million dollars already available, will accelerate development of the Rover nuclear rocket. This gives promise of some day providing a means for even more exciting and ambitious exploration of space, perhaps beyond the moon, perhaps to the very end of the solar system itself."*

That summer NASA began to fund a series of studies on manned missions to Mars that would use the experimental nuclear rocket known as *Rover*. It would later be called *NERVA* and to this day its derivatives are still considered mankind's best hope for sending humans to Mars in a short enough time frame to give them some chance of surviving the journey.

Several contractors engaged their brightest talents to investigate the task of sending humans to Mars. What they discovered was a daunting problem; many thought it might be impossible. It was known that the Earth passed Mars in space about once every 780 days and so the best time to launch a mission would be at one of those conjunctions. The problem is that the distance changes widely due to the varying distances between the Earth and the Sun and Mars and the Sun. About every 284 years the two planets get very close and about every 57 thousand years they become even closer. Obviously there was no way to wait for either of these latter close approachs (although one of the latter just passed in 2003 thus the opportunity has been lost for an almost uniquely short voyage.) As the mathematicians compiled the figures it seemed that a large amount of propellant would be necessary to effect even the remotest chance of getting to Mars in a manageable period of time. Using nuclear engines it might be possible to send a crew to Mars and bring them home again in as little as 400 days but a host of obstacles would have to be overcome.

The shorter the mission, the faster the spacecraft arrives back at Earth. Re-entry into the Earth's atmosphere could be as fast as 33,000 miles per hour. Clearly this was going to require some breakthrough engineering. Then there was the radiation hazard. America's first satellite in 1958 (Explorer 1) had discovered radiation belts around the Earth. Named after the designer of the experiment, these Van Allen belts clearly delineated the magnitude of solar radiation. The Earth was protected by its own intense magnetic field, but how would a crew aboard a spacecraft survive once they left the cocoon of magnetic force?

The studies revealed that the sun has its own weather. It has a cycle, and every eleven years or so it spews out high energy particles into space. These protons can be charged as high as 20 billion electron volts. On a mission to Mars the chances of encountering a blast of deadly solar radiation were one in ten. So why not shield the crew? It was soon realized that the shield would have to be as high as 300 pounds per square foot to keep the crew's exposure down to 30 REMs per hour. The accepted industrial level for exposure is only 5 REM per *year*. These early studies required as much as 70 *tons* of metal shielding for a seven man crew and perhaps as little as 23 tons if you were going to take a chance that no solar flares happened during the voyage. What was not known, at that time, was that Mars has virtually no magnetic field—therefore, while on the surface of Mars the crew would only be shielded by the thin Martian atmosphere. Today's best estimates suggest that direct overhead sunlight on Mars is equivalent to about 35 pounds per square foot of aluminum shielding. Metal shielding has been almost entirely discarded in today's studies in favor of water or polyethylene. Meanwhile it was obvious that more data was needed before humans could make the trip. It would first be necessary to send robots.

Over the next several years the race was set firmly between the Soviets and the United States as each country fought to prove their technological superiority. Just over five years after launching Sputnik 1, Korolev was ready to begin the era of Mars exploration with the launch of an unmanned probe later dubbed Mars-1. It was not in fact the first Mars probe launched by the Soviets but its three predecessors had not made it further than Earth orbit and so in keeping with the Soviet tradition of not naming failures, the first successful man made object to approach Mars was called Mars-1.

A triumph of ingenuity, Mars-1 included a pressurized camera module, solar panels for power, thermal control systems, high and low gain antennas for transmitting data, a science package, a solar tracker, attitude control jets, and triple redundancy transmitters. The entire craft weighed in at 2000 pounds. Sadly the Mars-1 probe would be the first to arrive in Mars-space but it would also lose contact with Earth despite several back-up radios. No pictures of Mars would be relayed home. It was, however, a tremendous accomplishment for 1962. Mars-1 proved that the technology existed for humans to finally reveal Mars' secrets—and those of any potential Martians.

Within two years the United States was ready to compete with the Russians at Martian exploration. Building on a string of successful earth satellites the team at the Jet Propulsion Laboratory in Pasadena were ready for the long-haul to Mars. On November 5th 1964 a 500 pound marvel of electronic engineering called *Mariner 3* sat on the launch pad in Florida. The launch was to take place aboard an uprated *Atlas* missile, developed by Convair of San Diego. The Atlas would have a second stage known as *Agena*. Unfortunately the shroud around the spacecraft did not deploy and the spacecraft was

Mariner 4 departs for Mars atop an Atlas Agena missile

subsequently too heavy to make the critical Martian trajectory. It would never make it to Mars. Just over three weeks later, on November 28th, Mariner 3's sister ship launched flawlessly and finally began America's multi-million mile journey to Mars.

228 days after launch *Mariner 4* flew blazingly quickly past the red planet while the controllers back on Earth awaited news. It would take more than a week for the robot to reveal its secrets since its transmitter was only capable of a transmission rate of just over eight bits per second—but it was worth the wait. For the first time in human history the surface of Mars was about to be revealed in detail, through the miracle of technology known as a fax-camera. Mariner 4 returned 21 complete photographs of Mars and a final incomplete 22nd picture.

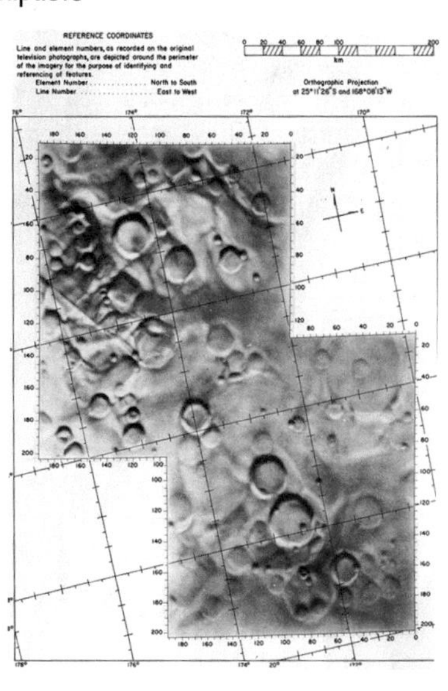

Mariner 4 pictures taken in 1965

A trend would begin with Mariner 4—a trend of perpetual Martian surprises. Few people expected to see a battered and cratered surface on the red planet but most of Mariner 4's pictures showed just that—it looked suspiciously like the Earth's moon. There was no sign of water, or canals, or ancient cities or fifteen foot tall green warriors or giant guns for launching war-ships. Mars looked sterile and dead. It was a tremendous disappointment for many people but it was just the beginning of an epic voyage. In the years ahead Mars would confound and startle and it would do it without once showing any signs of indigenous life. Mars looked almost exactly as Richard Proctor had described it nearly a hundred years earlier but, despite its superficially arid face, Mars' allure would attract dozens of surrogate explorers over the next four and a half decades. The exploration had barely begun.

Historical Mars Missions

Mission, Country, Launch Date,	*Purpose, Results*
[Unnamed], USSR, 10/10/60,	Mars flyby, did not reach Earth orbit
[Unnamed], USSR, 10/14/60,	Mars flyby, did not reach Earth orbit
[Unnamed], USSR, 10/24/62,	Mars flyby, achieved Earth orbit only
Mars 1, USSR, 11/1/62,	Mars flyby, radio failed at 65.9 million miles
[Unnamed], USSR, 11/4/62,	Mars flyby, achieved Earth orbit only
Mariner 3, U.S., 11/5/64,	Mars flyby, shroud failed to jettison
Mariner 4, U.S.11/28/64,	first successful Mars flyby 7/14/65, returned 21 photos
Zond 2, USSR, 11/30/64,	Mars flyby, passed Mars but radio failed, returned no planetary data
Mariner 6, U.S., 2/24/69,	Mars flyby 7/31/69, returned 75 photos
M-69, USSR, 3/27/69	Mars Orbiter didn't reach Earth orbit
Mariner 7, U.S., 3/27/69,	Mars flyby 8/5/69, returned 126 photos
M-69-2, USSR, 4/2/69	Mars Orbiter didn't reach Earth orbit
Mariner 8, U.S., 5/8/71,	Mars orbiter, failed during launch
Kosmos 419, USSR, 5/10/71,	Mars lander, achieved Earth orbit only
M-71S/Mars 2, USSR, 5/19/71,	Mars orbiter/lander arrived 11/27/71, no useful data, lander destroyed
M-71S/Mars 3, USSR, 5/28/71,	Mars orbiter/lander, arrived 12/3/71, returned one partial photo

Mission, Country, Launch Date,	*Purpose, Results*
Mariner 9, U.S., 5/30/71,	Mars orbiter, in orbit 11/13/71 to 10/27/72, returned 7,329 photos
Mars 4, USSR, 7/21/73,	failed Mars orbiter, flew past Mars 2/10/74
Mars 5, USSR, 7/25/73,	Mars orbiter, arrived 2/12/74, lasted a few days
Mars 6, USSR, 8/5/73,	Mars orbiter/lander, arrived 3/12/74, little data return
Mars 7, USSR, 8/9/73,	Mars orbiter/lander, arrived 3/9/74, little data return
Viking 1, U.S., 8/20/75,	Mars orbiter/lander, orbit 6/19/76-1980, lander 7/20/76-1982
Viking 2, U.S., 9/9/75,	Mars orbiter/lander, orbit 8/7/76-1987, lander 9/3/76-1980; combined, the Vikings returned 50,000+ photos
Phobos 1, USSR, 7/7/88,	Mars/Phobos orbiter/lander, lost 8/89 en route to Mars
Phobos 2, USSR, 7/12/88,	Mars/Phobos orbiter/lander, lost 3/89 near Phobos
Mars Observer, U.S., 9/25/92,	lost just before Mars arrival 8/21/93
Mars Global Surveyor, U.S., 11/7/96,	Mars orbiter, arrived 9/12/97, currently conducting prime mapping mission
Mars 96, Russia, 11/16/96,	orbiter/lander, launch vehicle failed
Mars Pathfinder, U.S., 12/4/96,	Mars lander and rover, landed 7/4/97, last transmission 9/27/97
Nozomi (Planet-B), Japan, 7/4/98,	Mars orbiter, failed to reach Mars orbit due to a malfunctioning valve, currently in orbit around the Sun;
Mars Climate Orbiter, U.S., 12/11/98;	lost on arrival at Mars 9/23/99
Polar Lander/ Deep Space 2, U.S., 1/3/99,	lander/descent probes; lander and probes lost on arrival 12/3/99.
Mars Odyssey, U.S, 3/7/01,	Mars orbiter, arrived 10/24/01; currently conducting prime mission studying global composition, ground ice, thermal imaging
Mars Express / Beagle 2, Esa, 6/2/03,	Mars orbiter/lander Entered orbit 12/25/03; Landing failed 12/25/03
MER U.S., 6/10/03, 7/7/03	Mars lander/rovers, Landing 1/4/04 (Spirit) 1/25/04 (Opportunity), as of July 2005 the rovers continue to conduct geology experiments.

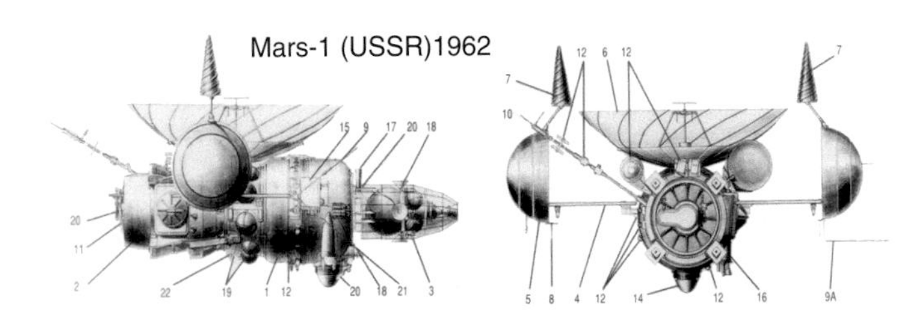

1. Pressurized orbital module
2. Special pressurized module (photo-module)
3. Correcting brake engine installation
4. Solar panel
5. Thermal control system radiators
6. High gain parabolic antenna
7. Low gain antenna
8. Low gain antenna
9. Meter wave-band transmitter antenna
9A. Meter wave-band receiver antenna
10. Omni-directional emergency radio antenna
11. Photo/TV and planet tracker portholes
12. Science instruments sensor
14. Precision solar and star tracker
15. Contingency radio link
16. Continuous solar tracker
17. Parabolic antenna Earth tracking sensor
18. Attitude control system nozzles
19. Attitude control system compressed gas tanks
20. Attitude sensor shutters
21. Non-precision sun tracker
22. Sun tracker

Total mass of object, KG	910
Mass radio instrumentation, KG	160
Mass Correcting Brake Engine, KG	68

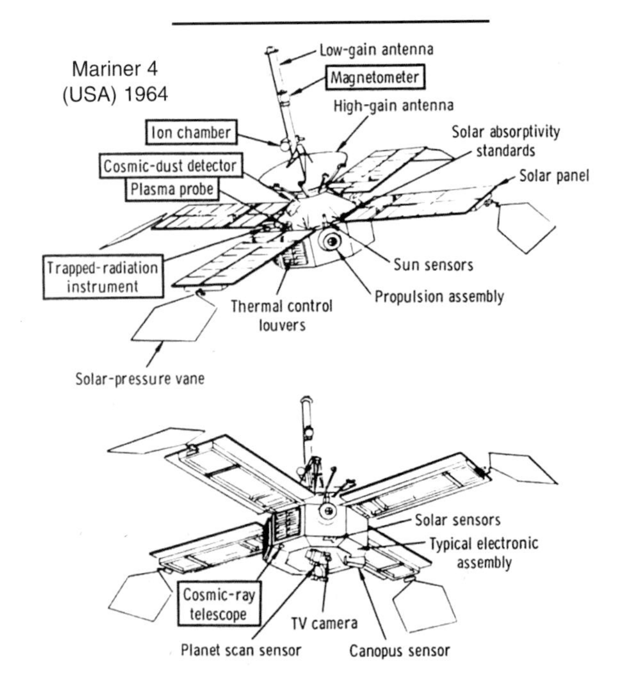

Mariner 6 & 7 (USA) 1969

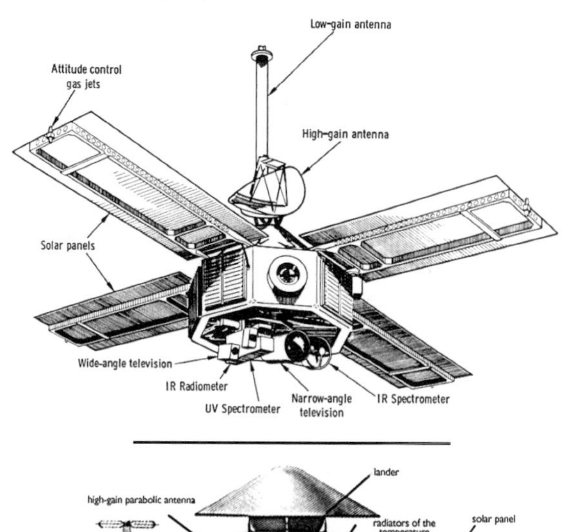
Low-gain antenna
Attitude control gas jets
High-gain antenna
Solar panels
Wide-angle television
IR Radiometer
UV Spectrometer
Narrow-angle television
IR Spectrometer

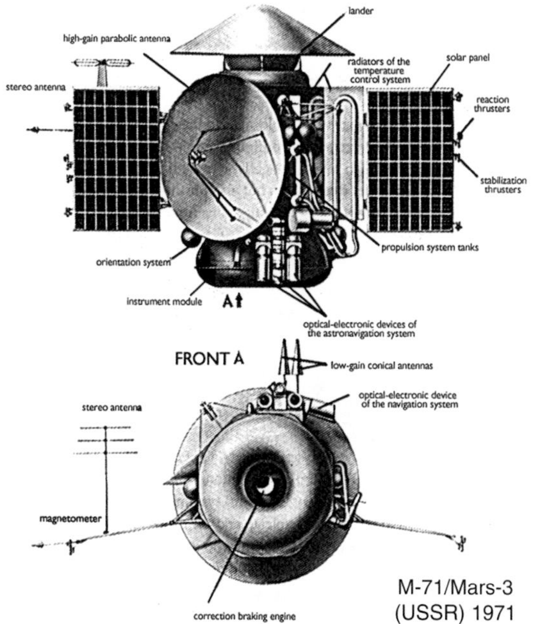
lander
high-gain parabolic antenna
radiators of the temperature control system
solar panel
stereo antenna
reaction thrusters
stabilization thrusters
propulsion system tanks
orientation system
instrument module
A
optical-electronic devices of the astronavigation system
FRONT A
low-gain conical antennas
optical-electronic device of the navigation system
stereo antenna
magnetometer
correction braking engine

M-71/Mars-3
(USSR) 1971

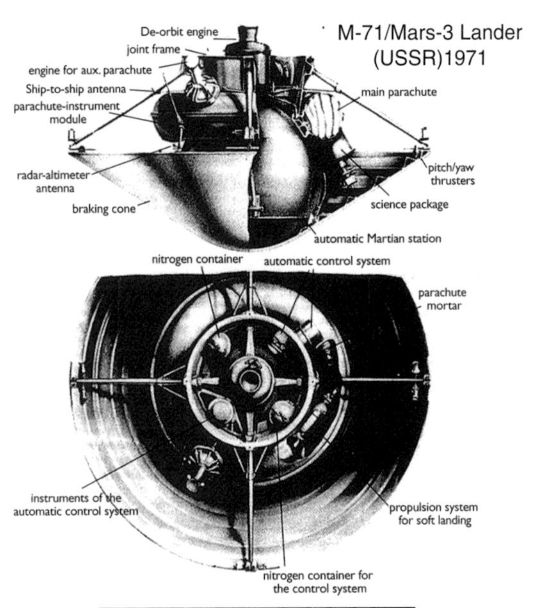

M-71/Mars-3 Lander (USSR)1971
De-orbit engine
joint frame
engine for aux. parachute
Ship-to-ship antenna
parachute-instrument module
main parachute
radar-altimeter antenna
pitch/yaw thrusters
braking cone
science package
automatic Martian station
nitrogen container
automatic control system
parachute mortar
instruments of the automatic control system
propulsion system for soft landing
nitrogen container for the control system

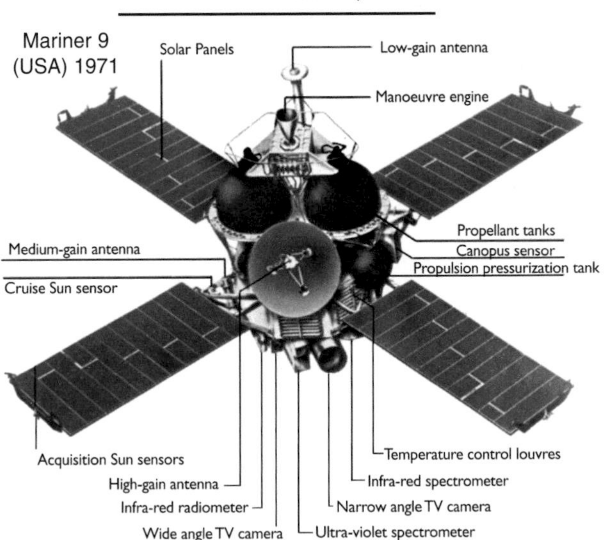

Mariner 9 (USA) 1971
Solar Panels
Low-gain antenna
Manoeuvre engine
Propellant tanks
Canopus sensor
Propulsion pressurization tank
Medium-gain antenna
Cruise Sun sensor
Acquisition Sun sensors
High-gain antenna
Infra-red radiometer
Wide angle TV camera
Temperature control louvres
Infra-red spectrometer
Narrow angle TV camera
Ultra-violet spectrometer

Viking 1 & 2
(USA)1976

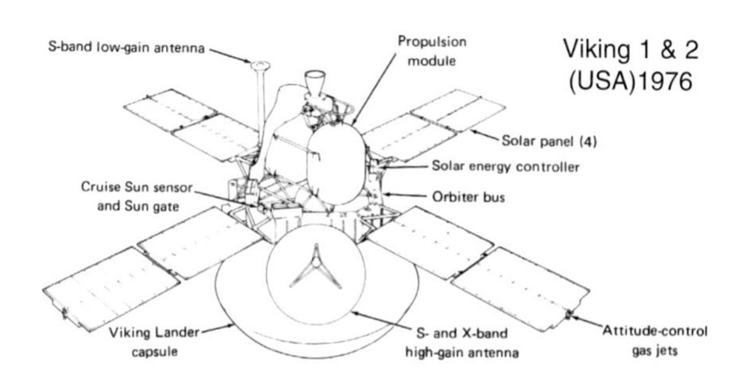
S-band low-gain antenna
Propulsion module
Solar panel (4)
Solar energy controller
Cruise Sun sensor and Sun gate
Orbiter bus
Viking Lander capsule
S- and X-band high-gain antenna
Attitude-control gas jets

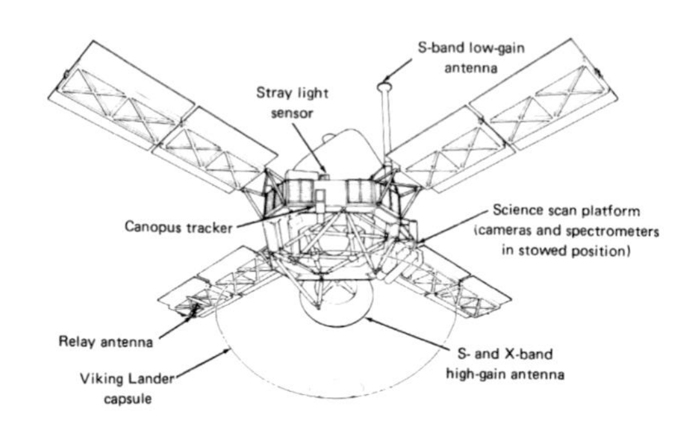
S-band low-gain antenna
Stray light sensor
Canopus tracker
Science scan platform (cameras and spectrometers in stowed position)
Relay antenna
Viking Lander capsule
S- and X-band high-gain antenna

VIKING LANDED SCIENCE CONFIGURATION

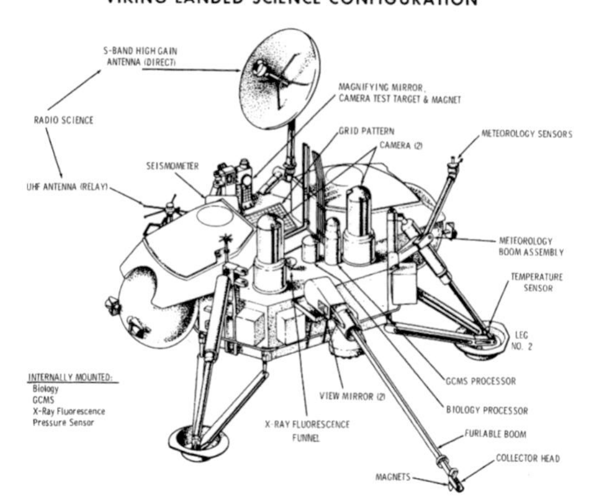
S-BAND HIGH GAIN ANTENNA (DIRECT)
RADIO SCIENCE
UHF ANTENNA (RELAY)
SEISMOMETER
MAGNIFYING MIRROR, CAMERA TEST TARGET & MAGNET
GRID PATTERN
CAMERA (2)
METEOROLOGY SENSORS
METEOROLOGY BOOM ASSEMBLY
TEMPERATURE SENSOR
LEG NO. 2
GCMS PROCESSOR
BIOLOGY PROCESSOR
FURLABLE BOOM
COLLECTOR HEAD
MAGNETS
VIEW MIRROR (2)
X-RAY FLUORESCENCE FUNNEL
INTERNALLY MOUNTED:
Biology
GCMS
X-Ray Fluorescence
Pressure Sensor

Mars Global Surveyor (USA) 1996

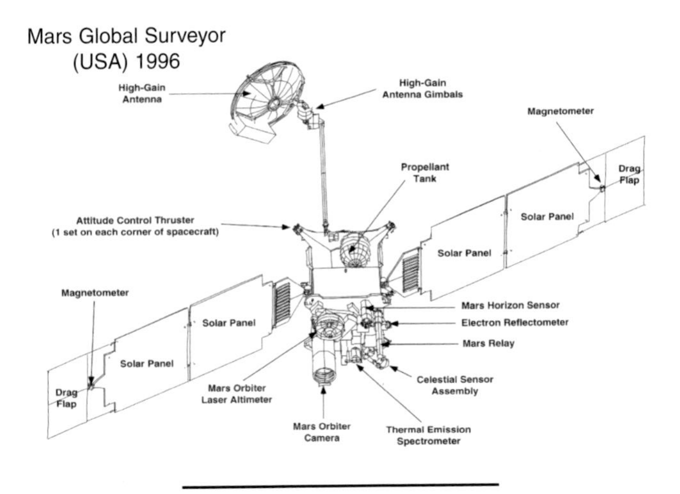

Mars Pathfinder/Sojourner (USA) 1997

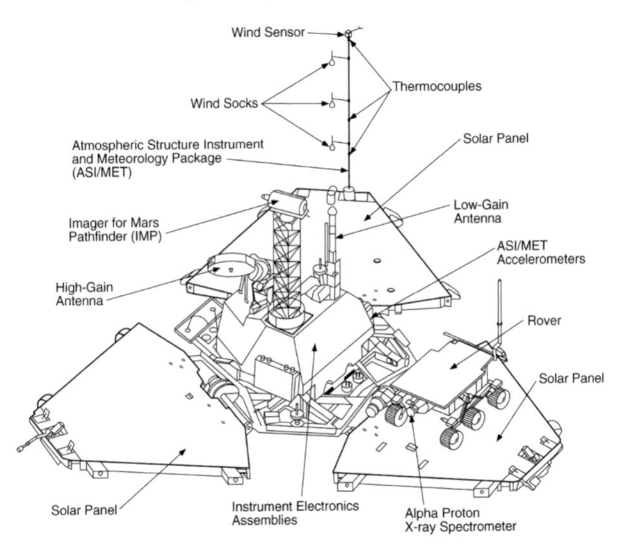

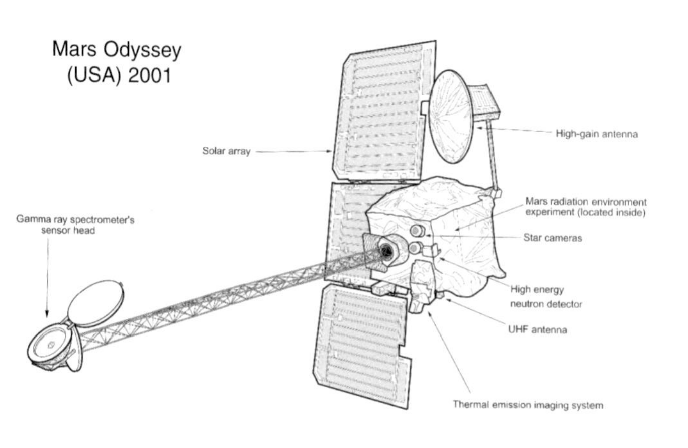
Mars Odyssey
(USA) 2001
Solar array
High-gain antenna
Mars radiation environment
experiment (located inside)
Star cameras
Gamma ray spectrometer's
sensor head
High energy
neutron detector
UHF antenna
Thermal emission imaging system

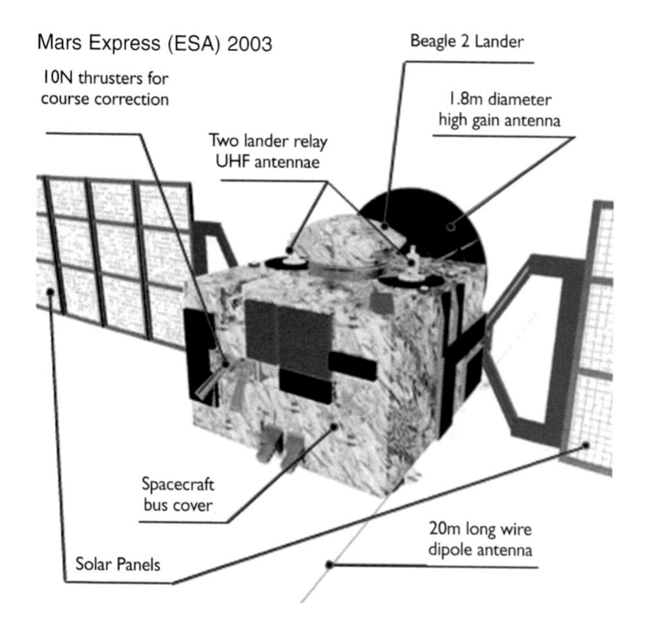
Mars Express (ESA) 2003
Beagle 2 Lander
10N thrusters for
course correction
1.8m diameter
high gain antenna
Two lander relay
UHF antennae
Spacecraft
bus cover
20m long wire
dipole antenna
Solar Panels

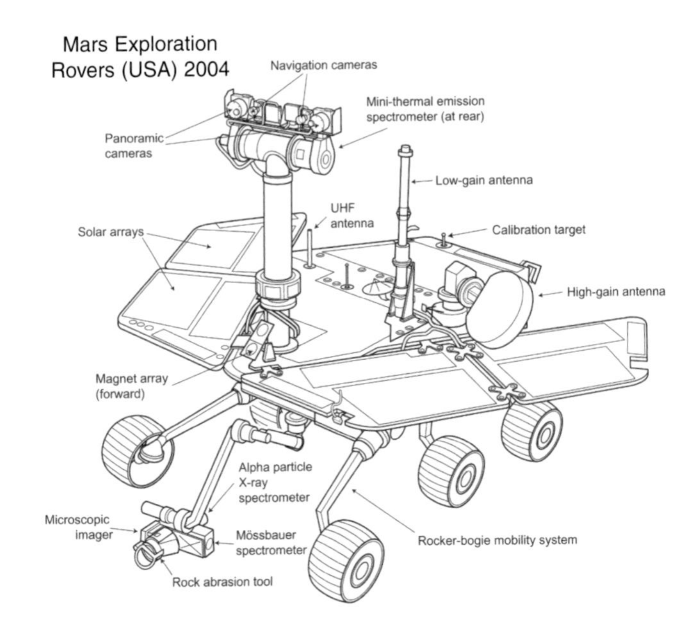

Beginning with the Soviet probe Mars-1, launched in 1962, there have been a total of fourteen successful Mars missions as of 2005. Mars-1 may not have completely fulfilled its mandate but it did prove conclusively that the propulsion, communications and guidance now existed to send a probe deep into the solar system. The 893kg vehicle was launched on a four-stage booster and completed the deployment of its solar panels, it then locked on to the Sun and began its journey to Mars. Shortly after departure it was seen that one of the orientation thrusters was leaking fuel. Ground controllers switched to gyros and kept the spacecraft powered-up by locking the panels onto the Sun. With no mid-course corrections possible the ship had no chance of reaching Mars but it was able to successfully transmit data about the space between Earth and Mars until it finally died at a distance of 106 million kilometers on March 21st 1963. As mentioned above, Mariner 4 would be launched in November 1964 and would become the first spacecraft to fly a completely successful mission to Mars with a flyby in the Spring of 1965. The next convenient window to attempt a Mars mission would not then open for another four years.

In 1969 the Soviets and the USA would both attempt to send two probes each to Mars. In the USSR they were called M-69 and in the USA they were called Mariner 6 and 7. Mariner 6 was nearly destroyed on the launch pad when the Atlas booster's main valves failed and the rocket began to deflate like a punctured balloon. Two daring ground-crew ran to the rocket and saved the day. Meanwhile, both Soviet vehicles would fail shortly after launch, while both NASA probes would go on to succeed with spectacular results. Arriving in the summer of 1969, flying high-energy trajectories which resulted in a short five month voyage, the Mariners would return over 200 high-resolution photographs of Mars. Both Mariners were the first planetary spacecraft capable of being re-programmed from the ground, this was a major breakthrough which was complimented by the much faster transmission rates available through the brand-new Goldstone 64-meter dish. The sharpest resolution pictures were able to reveal objects as small as 300 metres. The two Mariners would discover that the atmosphere of Mars was seemingly absent of Nitrogen and there was also no life-protecting ozone.

The next reasonable time to send probes to Mars was in 1971. This time the Soviets would send their latest spacecraft, designated M-71S. Once again, a pair would be sent, called Mars-2 and Mars-3. Both vehicles accomplished the first landings on Mars but unfortunately Mars-2 would fail on contact with the surface and its sister ship would only work for a few seconds, barely enough time to send back one extremely fuzzy photograph. The picture suggested that the lander may have either landed upside-down or been blown over by a world-girdling sand-storm that happened to have arrived just before the probes landed. Mars-3 would however succeed in placing the first Russian artificial satellite in Mars orbit. The lander would lose its ability to take pictures due to the coronal discharge casued by the hellish dust-storm while the orbiter would be deprived of returning any decent pictures because the entire surface was obscured by dust. However, a wealth of information was returned by Mars-3, such as the temperature and pressure on the surface, data on soil density, conductivity, permeability and reflectivity and information on Martian altimetry and water vapor.

The USA would also launch two probes in May of 1971, the first, Mariner 8, would fail on launch but Mariner 9 would fulfill its mission, arriving at Mars in November of that year. It would be the first craft to enter Mars orbit, arriving just ahead of Mars-3, and it would be able to dawdle long enough to wait for a clear sky. Mariner 9 would only need one course correction and would arrive within 50 kilometers of its target point, a remarkable accomplishment by any standards. On September 22nd, just two months before Mariner 9 and Mars-3 were to arrive, a white patch appeared over the Noachis region of Mars. In an uncanny real-life replay of a scene from Serviss's *Edison's Conquest of Mars*, the red planet seemed to be drawing a mantle around its secrets.

Earth's first emissaries were to be greeted by a Martian dust-storm of unprecedented proportions. While Mars-3 would linger a short time and return some good color pictures of the surface, the Mariner 9 probe was more tenacious and would stay until long after the dust had settled. Mariner 9 would completely re-write the book on Mars. Over 7000 pictures were taken from Mars orbit, small features were revealed for the first time. A massive canyon called Valles Marineris stretched thousands of miles and what looked like ancient river-beds were photographed in great detail. A cluster of objects were seen to emerge briefly above the dust storm on November 14th and showed that they were immense volcanoes, the largest was named Mount Olympus. The storm raged at wind-speeds of up to 180 kph and the surface temperature fluctuated by as much as 40 degrees Celsius. The storm was later seen as a blessing since it represented an unprecedented opportunity to study Mars in a totally different light.

The staggering amount of information returned by Mariner 9 set the stage for the first in a series of spectacular landing missions. Project *Viking* would be the most ambitious and expensive Mars missions to date. In keeping with previous attempts, two orbiter/landers would be launched at close intervals from the launch complex in Florida. The Vikings would be much larger than anything previously attempted by the United States. Viking would require three times as much fuel as Mariner 9, mostly to slow itself and the lander down so it could enter Mars orbit. The landers used the latest in small nuclear reactors to provide power. This surplus of power would help maintain the two landers during cold nights on the Martian surface. To send such large vehicles required a significant upgrade in the launch vehicle. Viking 1 and Viking 2 were both launched atop the Titan III-E rocket with Centaur upper stages. Viking 1 left Earth on August 20th 1975 and Viking 2 on September 9th. Shortly after launch both vehicles underwent significant course corrections since they had been launched to deliberately miss Mars by 67,000 km. This strange course of action was taken so that the upper stage of the booster would not accidentally hit Mars and violate the strict quarantine measures taken by the team at NASA. Viking was on its way to look for life and no one wanted to skew the results by trailing terrestrial organisms along by accident. On July 20th 1976, Viking 1 landed safely at a place called Chryse Planitia (the Plains of Gold), the name was originally derived from the term chosen almost a hundred years earlier by Schiaparelli. Meanwhile Viking 2 gently settled down at a place called Utopia Planitia on almost exactly the opposite side of the planet. Meanwhile the orbiters settled into Martian orbit and began work. Viking 2 would transmit data for another two years while Viking 1 would last over four years, finally expending its resources in August of 1980 after 1,489 orbits. Meanwhile the landers would fare almost as well with Lander 2 lasting for nearly four years and Lander 1 lasting almost six years. The copious wealth of data returned by

Viking would keep scientists busy for a generation. The most intriguing and controversial results were from the biology experiments aboard the two landers which both transmitted enigmatic and frustratingly inconclusive answers. The first really clear pictures from the surface of Mars showed a barren but starkly beautiful landscape very similar to the deserts of the American west. The experiments performed almost flawlessly, only the seismometer on Viking I failed. The soil was found to be extremely hostile to life. Less organic chemicals were present than on the Moon. Weather patterns were recorded and nitrogen was finally found in the atmosphere. Water vapor was quite abundant in the summer near the north pole and permafrost was suspected in substantial quantities. In one final triumph the Vikings were able to provide support for Einstein's Theory of Relativity with an unprecedented degree of accuracy when their transmissions were delayed by passing near the Sun.

Two decades would pass before another successful mission would reach Mars. Launched once again by the United States' National Air & Space Administration, the spacecraft was called Mars Global Surveyor (MGS). The mission mandate was to provide an unprecedented high resolution map of the Martian surface. MGS was launched November 7th 1996 and took ten months to arrive at Mars. Using a new technique perfected on an earlier mission to Venus, MGS slowed itself by skipping in and out of the Martian atmosphere for over four months. This technique required less fuel (and thus less weight) and allowed the vehicle to be launched in a much cheaper rocket than that used by Viking. The spacecraft began its prime mapping mission in March 1999. It observed the planet from a low-altitude, nearly polar orbit. MGS completed its primary mission on January 31, 2001, and went into an extended mission phase. MGS has studied the entire Martian surface, atmosphere, and interior, and has returned more data about the red planet than all other Mars missions combined. MGS has taken pictures of gullies and debris flow features that suggest there may be current sources of liquid water, similar to an aquifer, at or near the surface of the planet. Magnetometer readings show that the planet's magnetic field is not globally generated in the planet's core, but is localized in particular areas of the crust.

Mars Pathfinder was originally designed as a technology demonstration of a way to deliver an instrumented lander and a free-ranging robotic rover to the surface of the red planet. Pathfinder not only accomplished this goal but also returned an unprecedented amount of data and outlived its primary design life. Mars Pathfinder used an innovative method of directly entering the Martian atmosphere, assisted by a parachute to slow its descent through the thin Martian atmosphere and a giant system of airbags to cushion the impact. The landing site, an ancient flood plain in Mars' northern hemisphere known as Ares Vallis, is among the rockiest parts of Mars. It was chosen because scientists believed it to be a relatively safe surface to land on and one which contained a wide variety of rocks deposited during an ancient catastrophic flood. From landing until the final data transmission on September 27, 1997, Mars Pathfinder returned more than 16,500 images from the lander and 550 images from the rover, as well as chemical analyses of rocks and soil and extensive data on winds and other weather factors. Findings from both the lander and the rover suggest that Mars was at one time in its past warm and wet.

Mars Odyssey is an orbiting spacecraft designed to determine the composition of the planet's surface, to detect water and shallow buried ice, and to study the radiation environment. Odyssey has collected images that will be used to identify the minerals present in the soils and rocks on the surface and to study small-scale geologic processes and landing site characteristics. By measuring the amount of hydrogen in the upper meter of soil across the whole planet, the spacecraft will help us understand how much water may be available for future exploration, as well as give us clues about the planet's climate history. The orbiter will also collect data on the radiation environment to help assess potential risks to any future human explorers, and can act as a communications relay for future Mars landers.

Mars Express is Europe's first spacecraft to the Red Planet. At launch, it carried seven instruments and a lander. The orbiter's instruments are now performing remote sensing of Mars from subsurface layers all the way up to free space. The lander, Beagle 2, would have performed on-the-spot investigations of the surface, using geochemical, exobiological and atmospheric parameters, but it was declared lost on 6 February 2004 after no signals were received. It was also designed to look for signs of past or present life.

Two powerful new Mars rovers, named Spirit and Opportunity are on the red planet. With far greater mobility than the 1997 Mars Pathfinder rover, these robotic explorers may trek as much as 40 meters across the surface in a day. Each rover carries a sophisticated set of instruments to search for evidence of liquid water that may have been present in the planet's past. The landing for each resembled that of the Pathfinder mission. A parachute deployed to slow the spacecraft, rockets fired to slow it further just before impact, and airbags inflated to cushion the landing. The Mars Rovers have outlived their life expectancy and as of October 2005 are still sending back spectacular pictures and revolutionary data from Mars.

More Mars missions are currently on the drawing boards and in 2004 President Bush announced a bold new initiative to ultimately send humans to Mars.

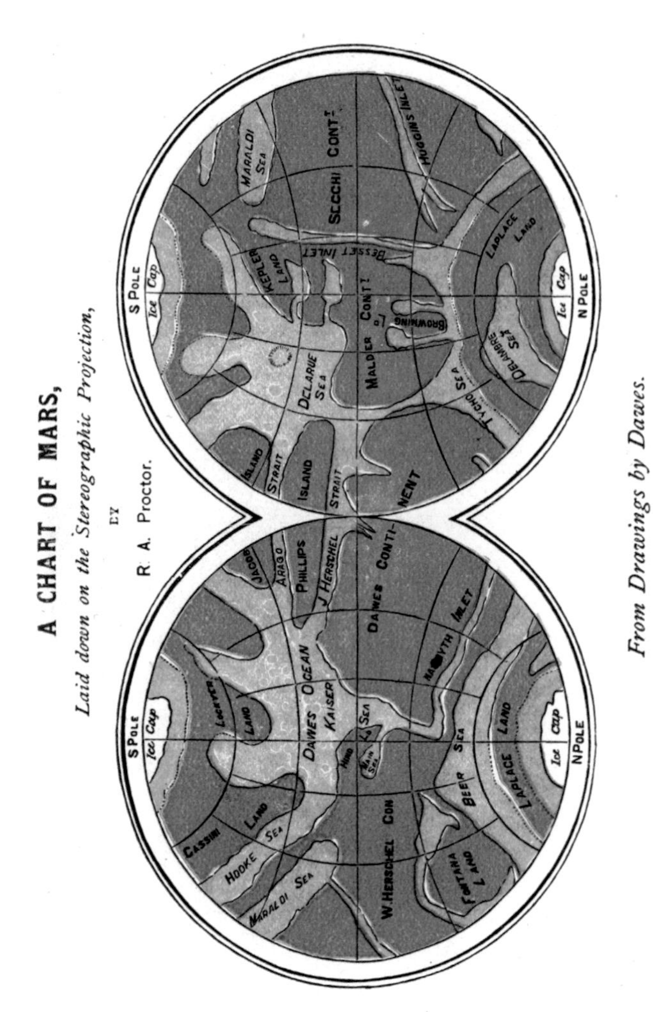

An 1871 map of Mars by Richard Proctor from drawings by William Dawes. Many astronomers still thought Mars had oceans at this time.

MARTE

secondo le osservazioni fatte col Tubo Equatoriale della Specola di Brera, Settembre 1877 _ Marzo 1878.

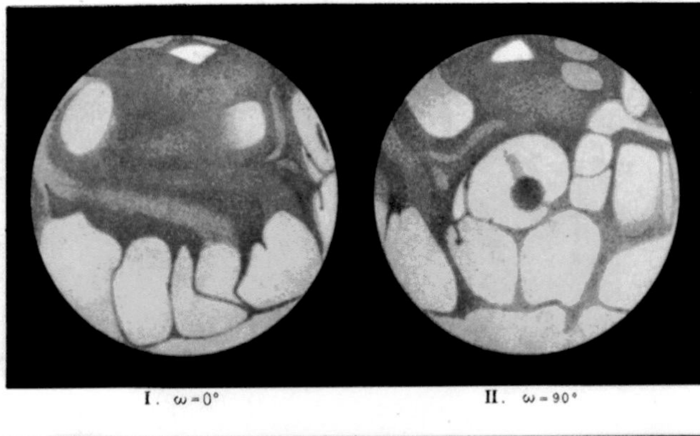

I. $\omega = 0°$ II. $\omega = 90°$

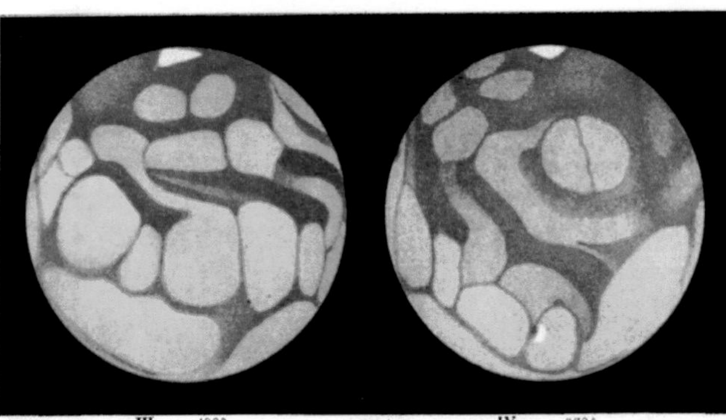

III. $\omega = 180°$ IV. $\omega = 270°$

Questi disegni sono projezioni ortografiche col centro nella longitudine areografica di 0°, 90°, 180° e 270° e nella latitudine areografica australe di 25°.

Drawings by Giovanni Schiaparelli of Mars during its close approachs of 1877-1888 and his famous map (opposite page). This was the first map to adopt the Latin names. Most of these names are still used today with the exception of the many thin lines which were thought to be channels of water. Later they were found to be optical illusions.

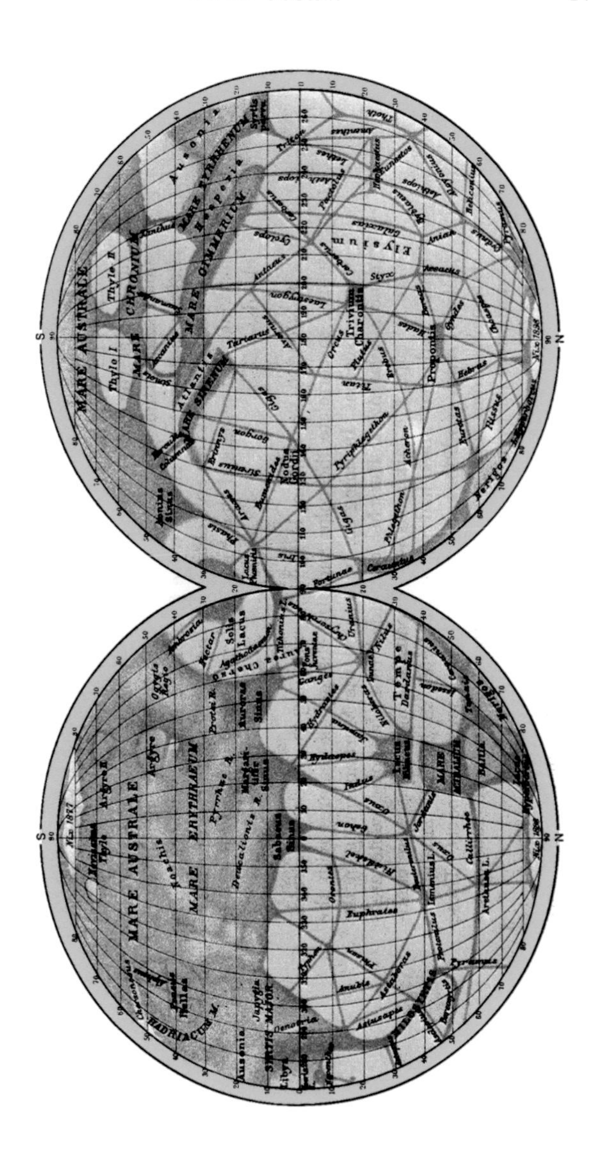
MARE AUSTRALE
MARE CHRONIUM
MARE CIMMERIUM
MARE TYRRHENUM
MARE SIRENUM
Elysium
Propontis
Trivium Charontis
Ausonia
Hesperia
Atlantis
Thyle I
Thyle II
MARE ERYTHRAEUM
Solis Lacus
Tempe
Sabaeus Sinus
SYRTIS MAJOR
HADRIACUM M.
Libya
Argyre
Deucalionis R.
Pyrrhae R.
Euphrates
Hiddekel
Gehon
Hydaspes
Ganges
Nilokeras
Styx
Cerberus
Gigas
Orcus
S
N

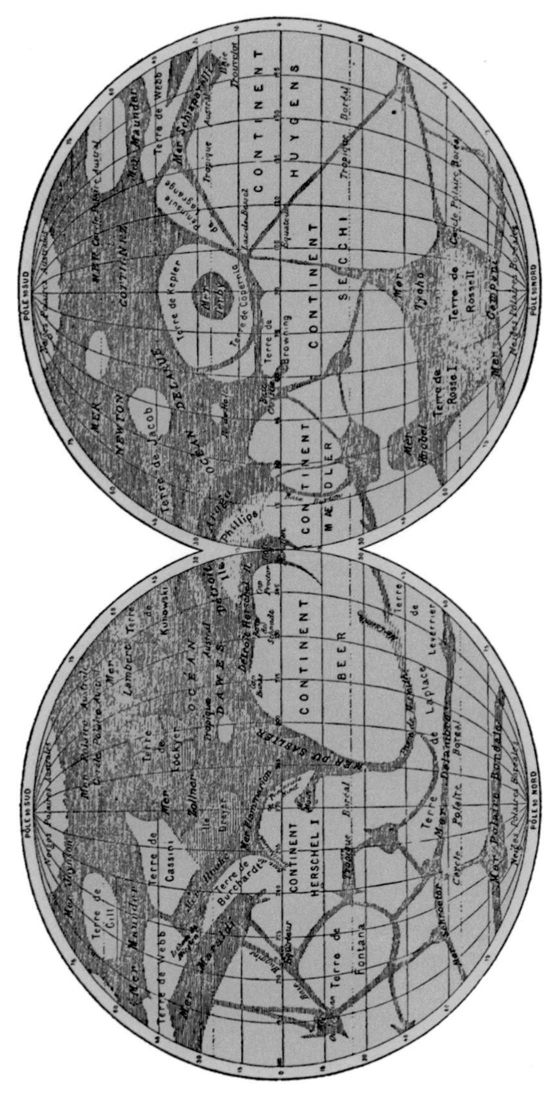

An 1892 map of Mars by Camille Flammarion
with some features named after famous astronomers

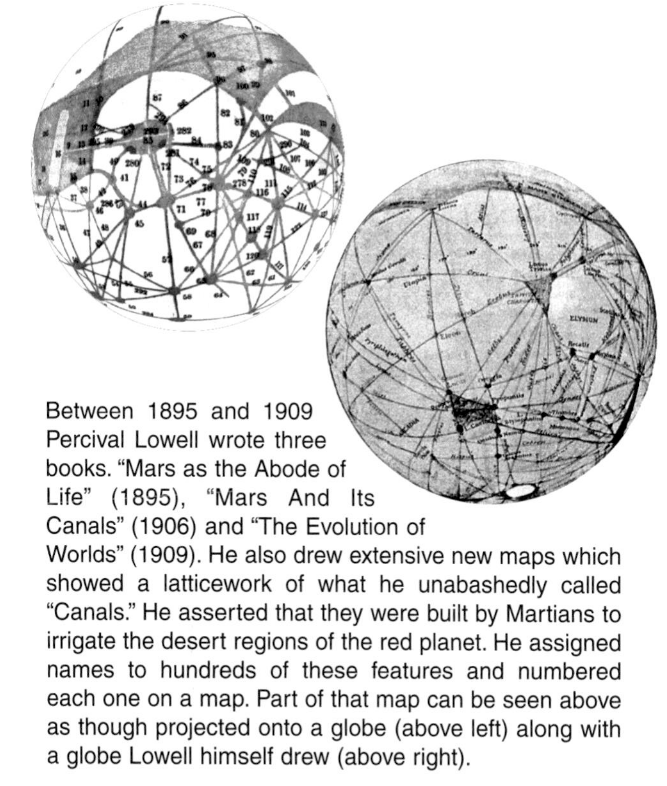

Between 1895 and 1909 Percival Lowell wrote three books. "Mars as the Abode of Life" (1895), "Mars And Its Canals" (1906) and "The Evolution of Worlds" (1909). He also drew extensive new maps which showed a latticework of what he unabashedly called "Canals." He asserted that they were built by Martians to irrigate the desert regions of the red planet. He assigned names to hundreds of these features and numbered each one on a map. Part of that map can be seen above as though projected onto a globe (above left) along with a globe Lowell himself drew (above right).

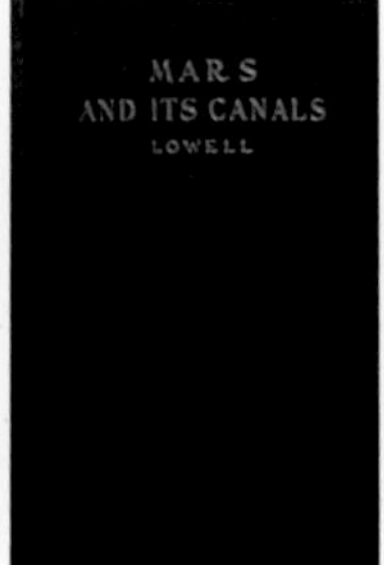

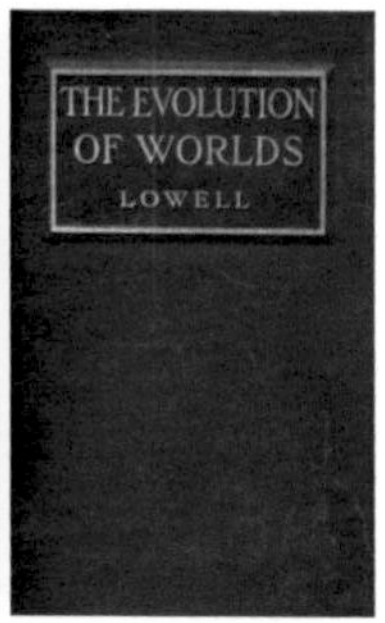

The discussion of life on Mars continued unabated during the first quarter of the 20th Century. Shown here are two science magazines published in 1915 and 1924 by Hugo Gernsback who would later go on to establish a market for science fiction magazines in the 1930's. While Gernsback cashed in on the public enthusiasm he also provided a forum for respectable scientists to publish fanciful ideas that would have been dismissed by the mainstream science journals.

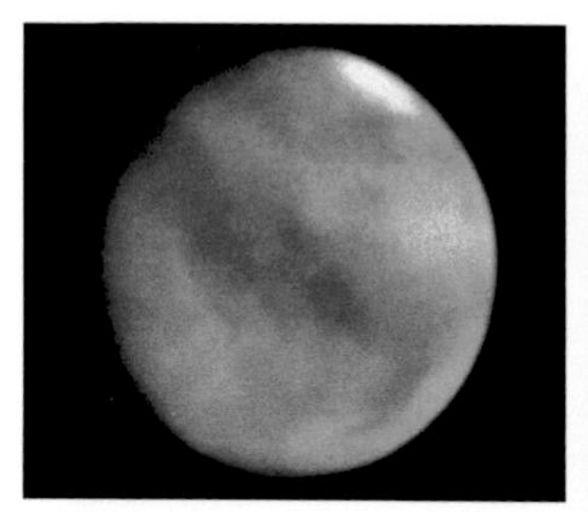

At the time the picture above was taken in 1956 it was one of the best images available of Mars. Taken with the Mount Wilson 60-inch reflector by R.B. Leighton of the California Institute of Technology.

Compare it to the image taken in 1969 by Mariner 7 (below) and the beautiful sequence taken by the Hubble Space Telescope (right) The Mariners were equipped with television cameras that used several filters to simulate color.

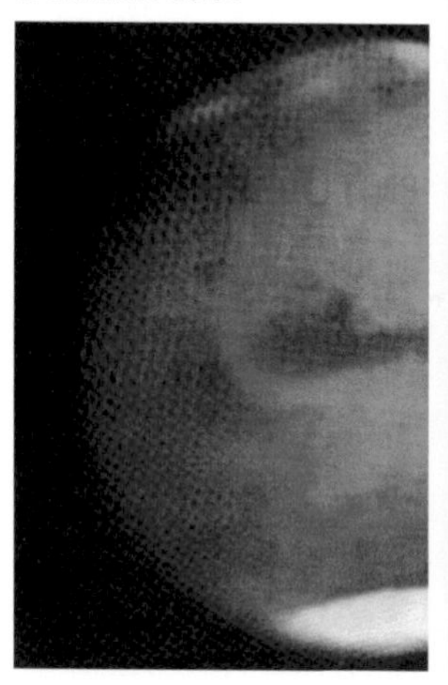

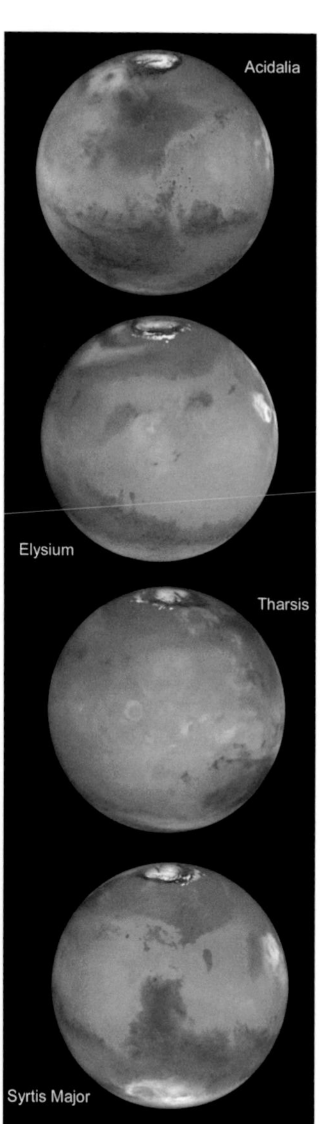

The 1950's was a fertile time for discussing trips to Mars. Wernher von Braun and artist Chesley Bonestell were recruited by Collier's Magazine for a series of influential articles (opposite). Bonestell provided a romantic yet precisely detailed vision of mankind's exploration of the planets. He is generally recognized as the greatest space artist to put brush to canvas. Von Braun published his template for a manned Mars mission in 1956 (inset). Originally published in German, it outlined in great detail the mechanisms necessary to reach Mars. Later he would work with Walt Disney to bring the Collier's articles to life in a series of animated movies.

(Courtesy Bonestell Space Art)

Collier's

APRIL 30, 1954 • FIFTEEN CENTS

Can We Get to Mars?

Is There Life on Mars?

SPECIAL REPORT

How Your Town C

AVOID

A Recession

•

8 Danger Signals To Watch For

•

10 Specific Steps To Prevent Troubl

This picture was the best color picture of Mars ever taken up to the summer of 1971. It was created using a system of filters on the television camera aboard Mariner 9. None of the Mariner spacecraft carried actual color cameras. Close-up color pictures of Mars would have to wait until the summer of 1976 when the super-sophisticated Viking spacecraft arrived.

A beautiful view of Mars taken on final approach by one of the Viking Orbiters launched in 1975 (below left)

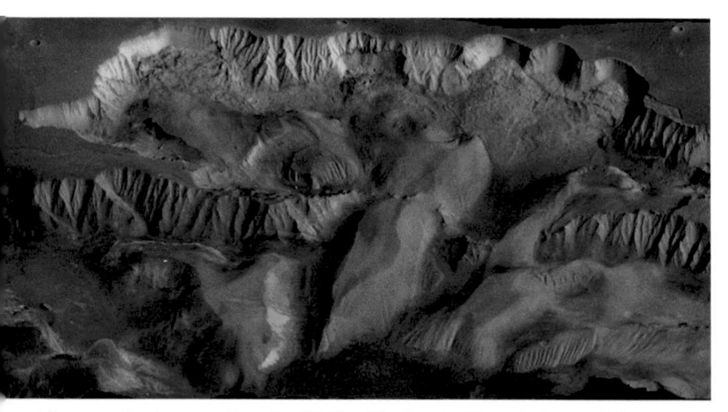

An oblique, color image of central Valles Marineris. The photograph is a composite of Viking high-resolution images in black and white and low-resolution images in color. Ophir Chasma on the north is approximately 300 km across and as deep as 10 km.

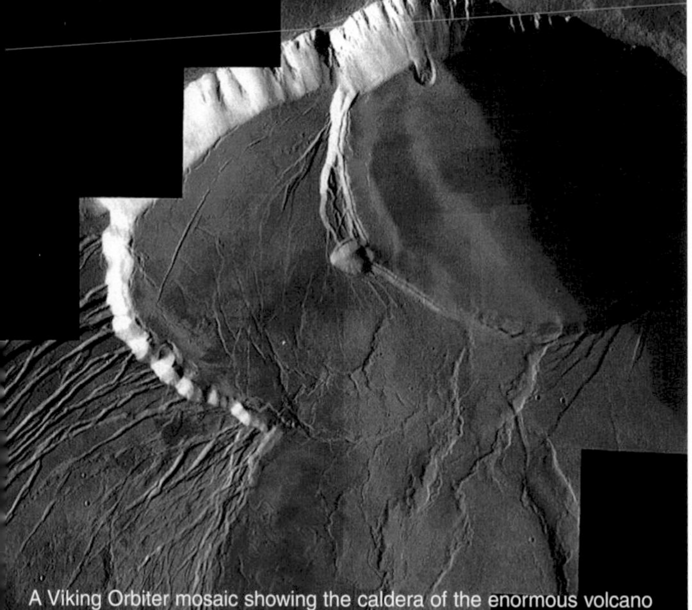

A Viking Orbiter mosaic showing the caldera of the enormous volcano Olympus Mons. The mountain stands nearly 27 km above the surrounding terrain. The image above is approximately 40 by 40 km.

The first full color picture from the surface of Mars returned by Viking 1. (above) The first color picture back from Viking 2 (below) The image was taken by camera 2 on 5 September 1976, two days after landing. The lander is at an angle of 8 degrees, so the horizon appears tilted. The colors of the rocks and soil are similar to those at the Viking 1 Lander site in Chryse Planitia.

Mosaic of the Valles Marineris hemisphere of Mars. (above right) The view is 2,500 kilometers from the surface of the planet. The mosaic is composed of 102 Viking Orbiter images of Mars. The center of the picture shows the entire Valles Marineris canyon system, over 3,000 kilometers long and up to 8 kilometers deep. The three Tharsis volcanoes, each about 25 kilometers high, are visible to the west. The Viking lander as it would have appeared after fully deploying its science package on the surface of Mars.

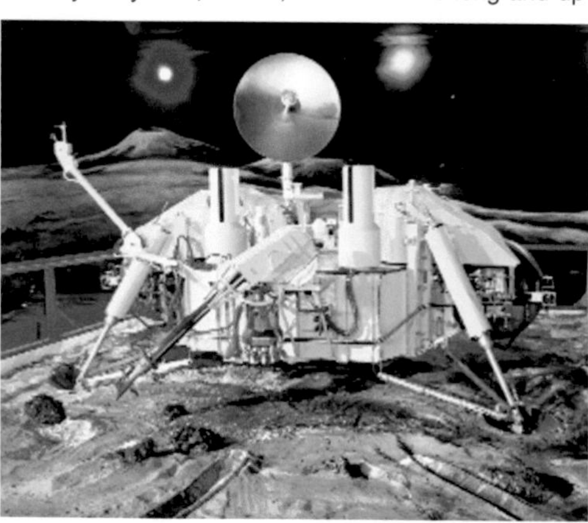

Viking 1 and 2 landed on opposite sides of Mars. Both robots were equipped with cameras which could send back panoramic images of their landing sites. The one above is the view from Camera 1/Viking 1 at Chryse Planitia. The image is false color. The large rock dubbed “Big Joe” can be seen almost dead-centre. Below is the Camera 2/Viking 2 landing site at Utopia Planitia. This panorama is a composite of true color and false color images. The cover from the sampling scoop can be seen discarded almost at dead-centre.

A view looking back across Viking Lander 2. Dark boulders are prominent against the yellowish-brown soil. Sunset on Mars (below)

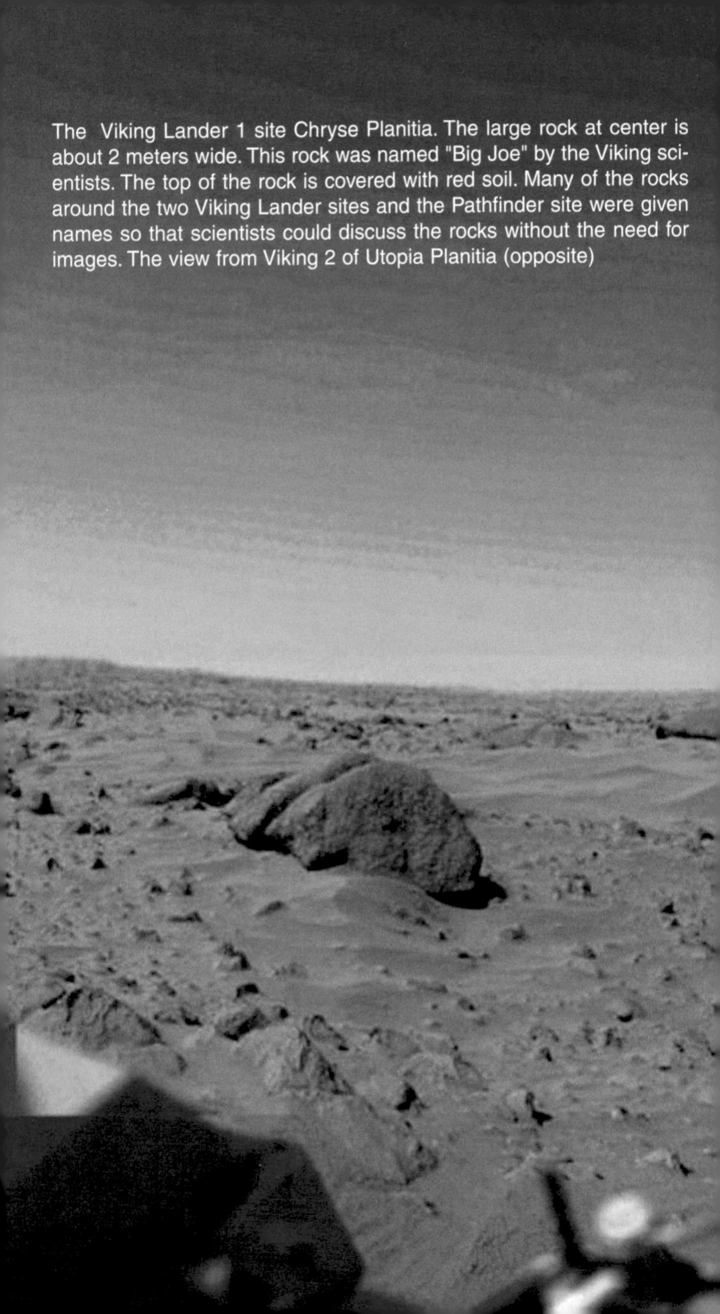

The Viking Lander 1 site Chryse Planitia. The large rock at center is about 2 meters wide. This rock was named "Big Joe" by the Viking scientists. The top of the rock is covered with red soil. Many of the rocks around the two Viking Lander sites and the Pathfinder site were given names so that scientists could discuss the rocks without the need for images. The view from Viking 2 of Utopia Planitia (opposite)

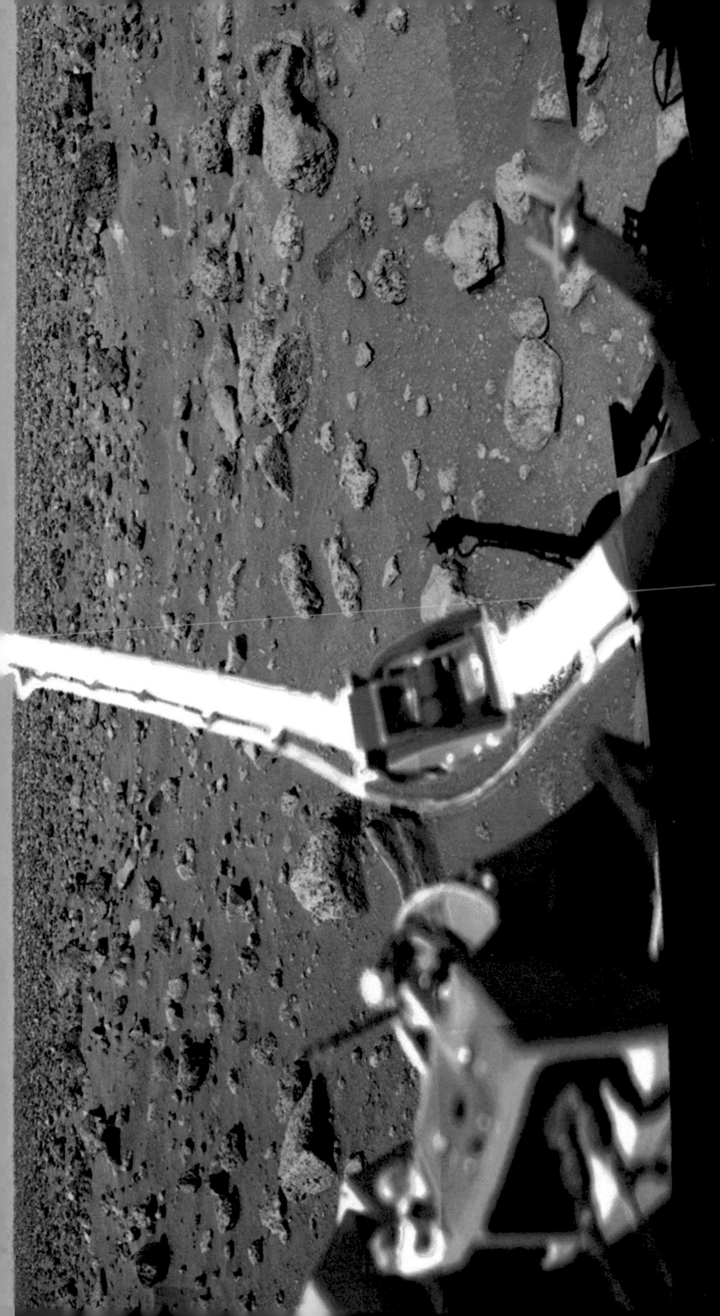

This image shows a thin layer of water ice frost on the martian surface at Utopia Planitia. It was taken by Viking 2 on 18 May 1979. It is speculated that dust particles in the atmosphere pick up tiny bits of water. When it gets cold enough for carbon dioxide to solidify, some of it attaches to the dust and ice and it falls to the surface.

Viking 1 Lander image of Chryse Planitia looking over the lander. The large white object at lower left and center, with the American flag on the side, is the radiothermal generator (RTG) cover. The high-gain S-band antenna is at upper right. The image was taken on 30 August 1976, a little over a month after landing.

Utopia Planitia as seen by Viking 2. The rounded rock in the center foreground is about 20cm wide. There are two trenches that were dug in the soil, the piece of hardware at right is the cover from the sampling scoop.

The Martian horizon, three kilometers distant, as seen by Viking 2 in September 1976. Viking's surface sampler is at left, the antenna for receiving commands from Earth at right.

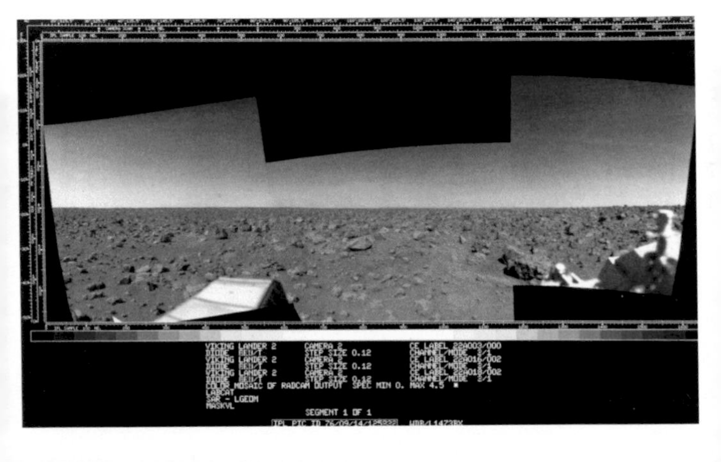

Three faces of Mars. The large shot at right was taken by compositing a series of over 100 photographs from the Viking orbiters.

The images above and below were taken by Mars Global Surveyor and show before and during a planetwide dust storm.

Another face reveals itself to the Viking cameras, this one a rock formation in a region called Cydonia (above). Some people today still think this represents proof of Martian life although when the feature was rephotographed by Mars Global Surveyor it was revealed to be nothing more than a trick of shadows (below).

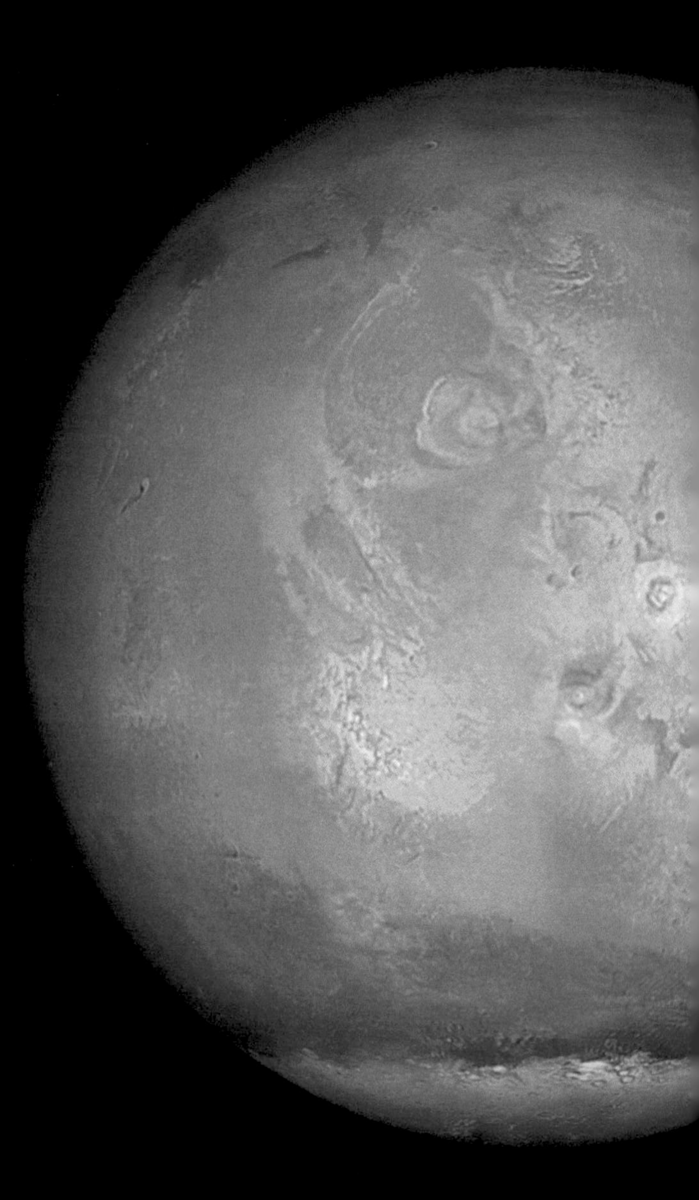

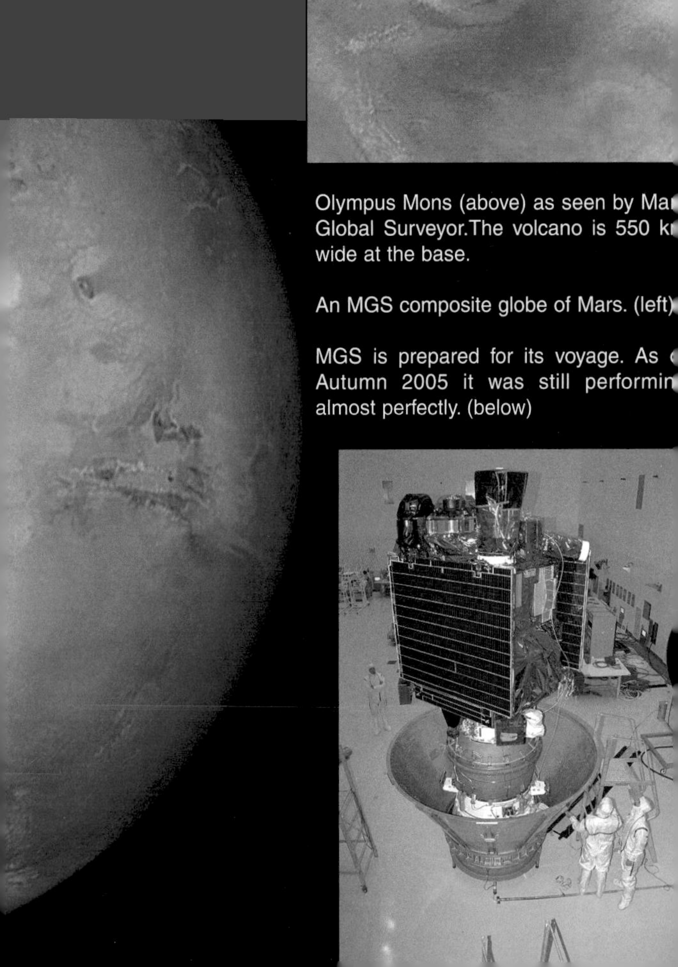

Olympus Mons (above) as seen by Ma
Global Surveyor. The volcano is 550 k
wide at the base.

An MGS composite globe of Mars. (left)

MGS is prepared for its voyage. As
Autumn 2005 it was still performin
almost perfectly. (below)

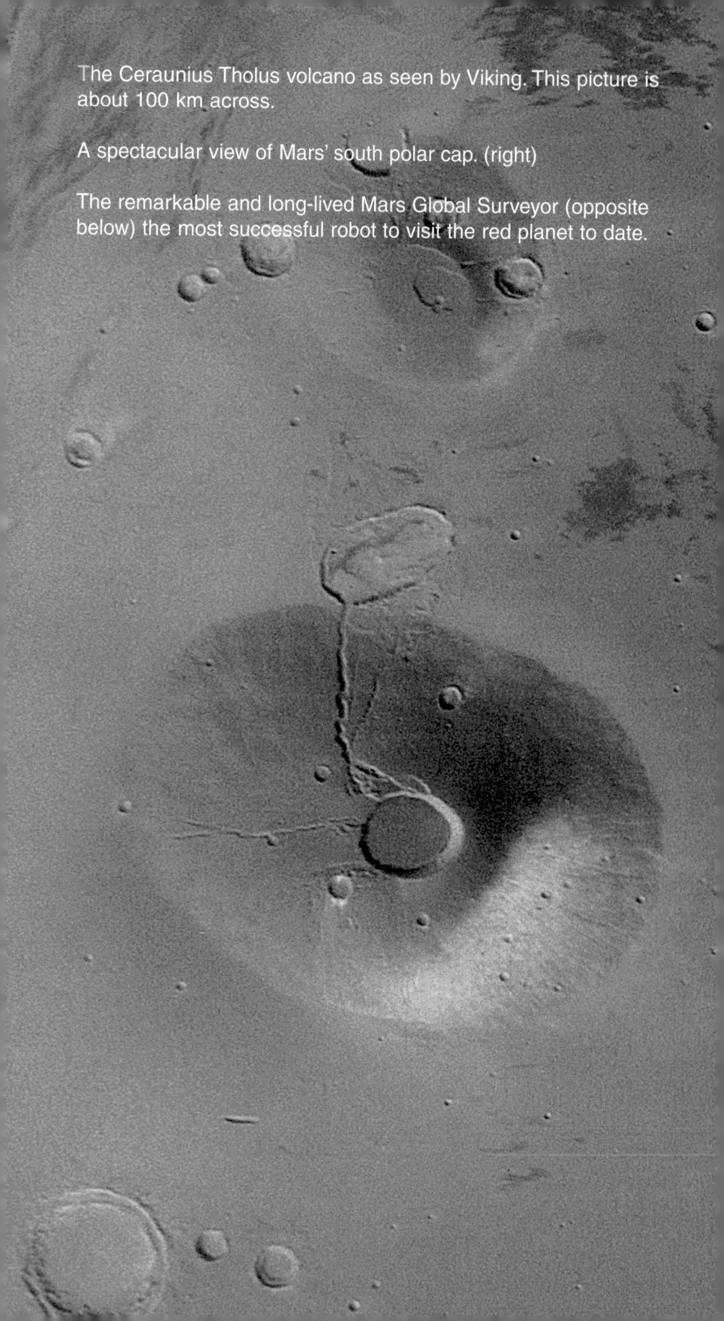

The Ceraunius Tholus volcano as seen by Viking. This picture is about 100 km across.

A spectacular view of Mars' south polar cap. (right)

The remarkable and long-lived Mars Global Surveyor (opposite below) the most successful robot to visit the red planet to date.

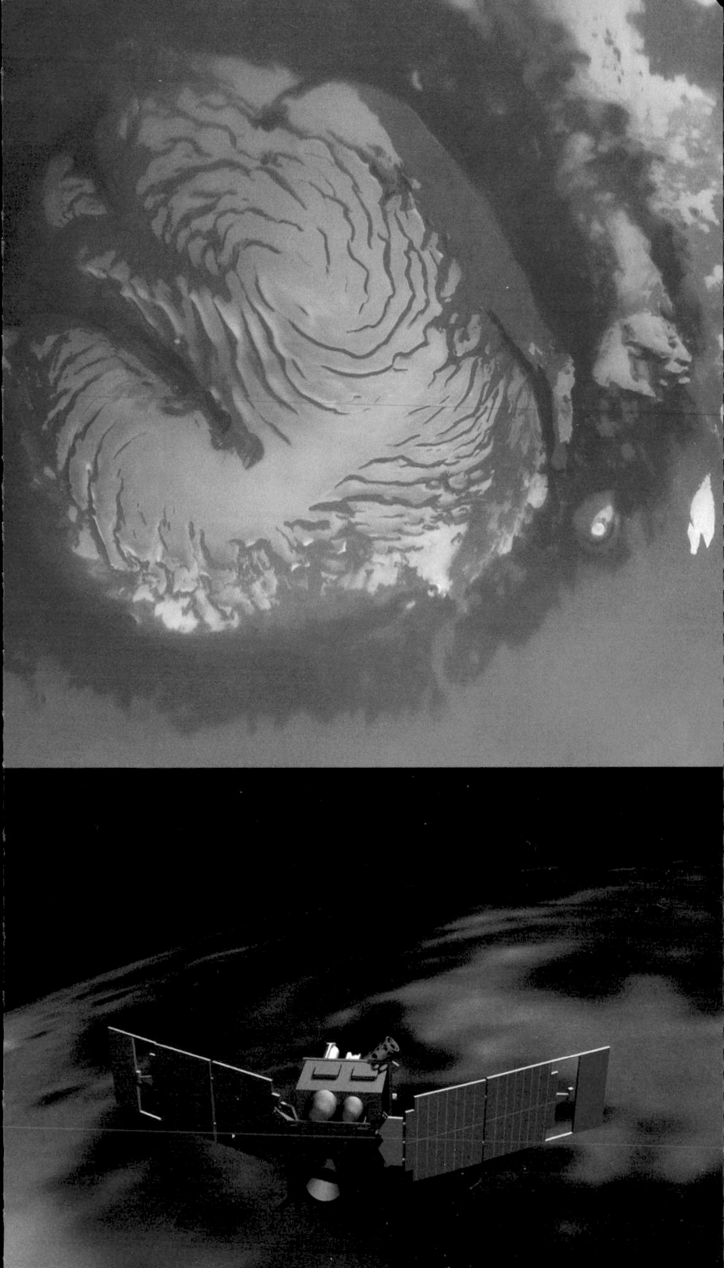

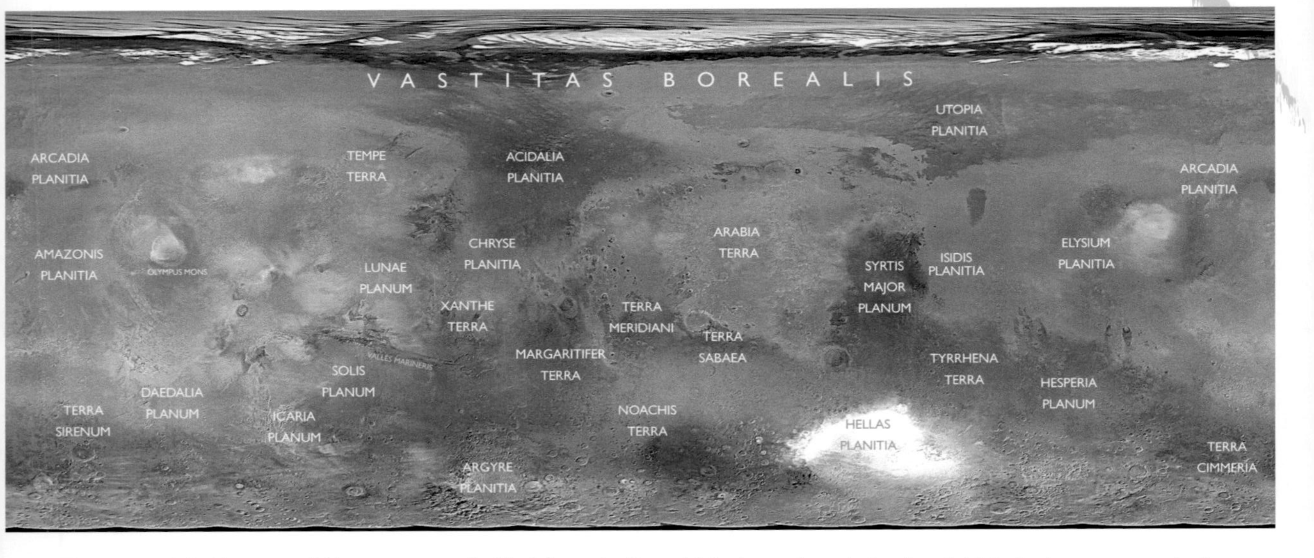

The latest global images of Mars are compiled in this projection which shows the whole planet. Main features are marked.

The Pathfinder Mars probe with its rover named Sojourner were launched to test a new landing system. The vehicle (inset) would plummet into Mars atmosphere and then using a series of untried mechanisms deliver the rover to the surface.

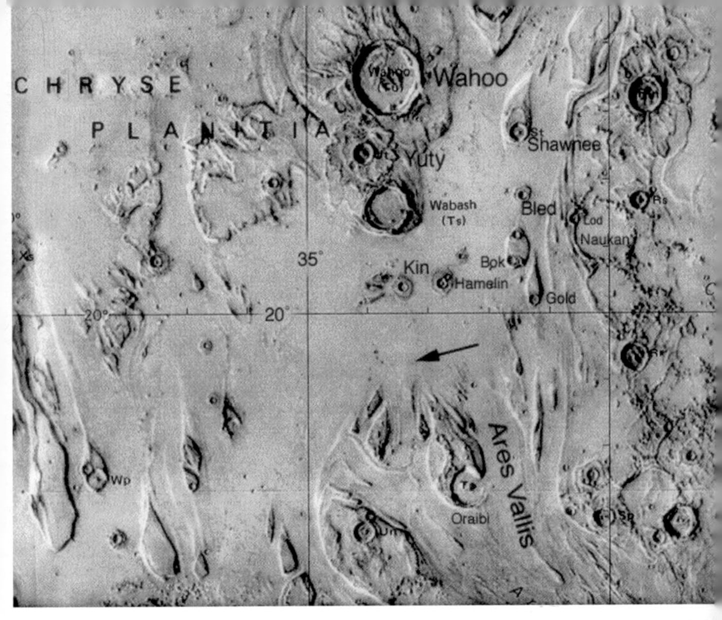

The landing site of Pathfinder/Sojourner (above). Engineers inspect a model the lander with its revolutionary airbag system deployed (below). After beir slowed by atmospheric friction and parachutes the airbags were inflated at tl last moment. The system was an advanced version of a system suggested 1960 which used an inflatable skirt below the lander (inset). It worked flawles ly and became the standard for future robotic American Mars landers.

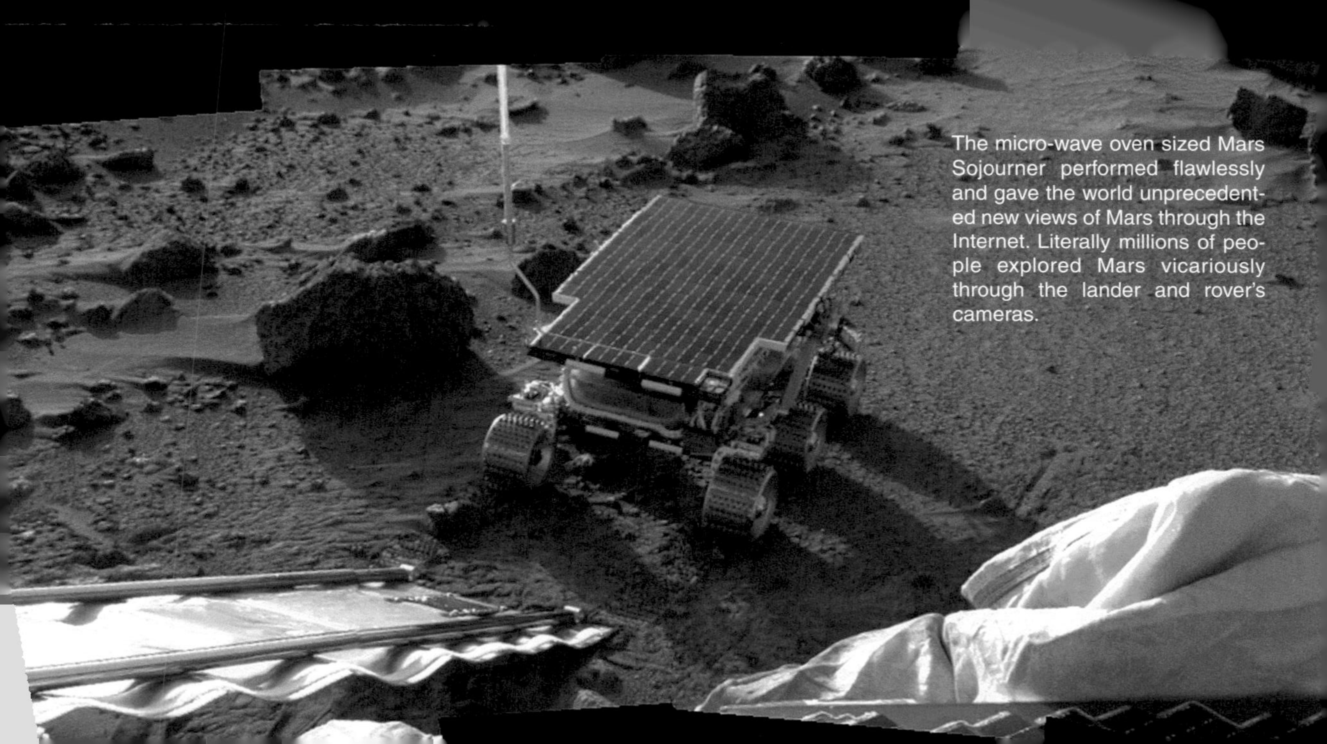

The micro-wave oven sized Mars Sojourner performed flawlessly and gave the world unprecedented new views of Mars through the Internet. Literally millions of people explored Mars vicariously through the lander and rover's cameras.

Astoundingly detailed images returned by the Pathfinder lander allowed scientists to study Mars in a detail previously unimagined. These two pages show a 360° panorama of the landing site. In keeping with a tradition started with the Viking landers, scientists assigned frivolous names to features to aid in their identification. The large rock at left was dubbed “Yogi” because it looked like a bear. In the distance (opposite page) the two hills were named “Twin Peaks.”

Plainly visible in the foreground are the two yellowish ramps which were unrolled to expedite the deployment of the Sojourner rover. The rover can be seen at far right "sniffing" the large rock called "Yogi". Tracks are clearly visible showing where the rover was commanded to dig up some dust in the hopes of revealing subsurface details.

Backshell

Parachute

Bounce marks

Spiri

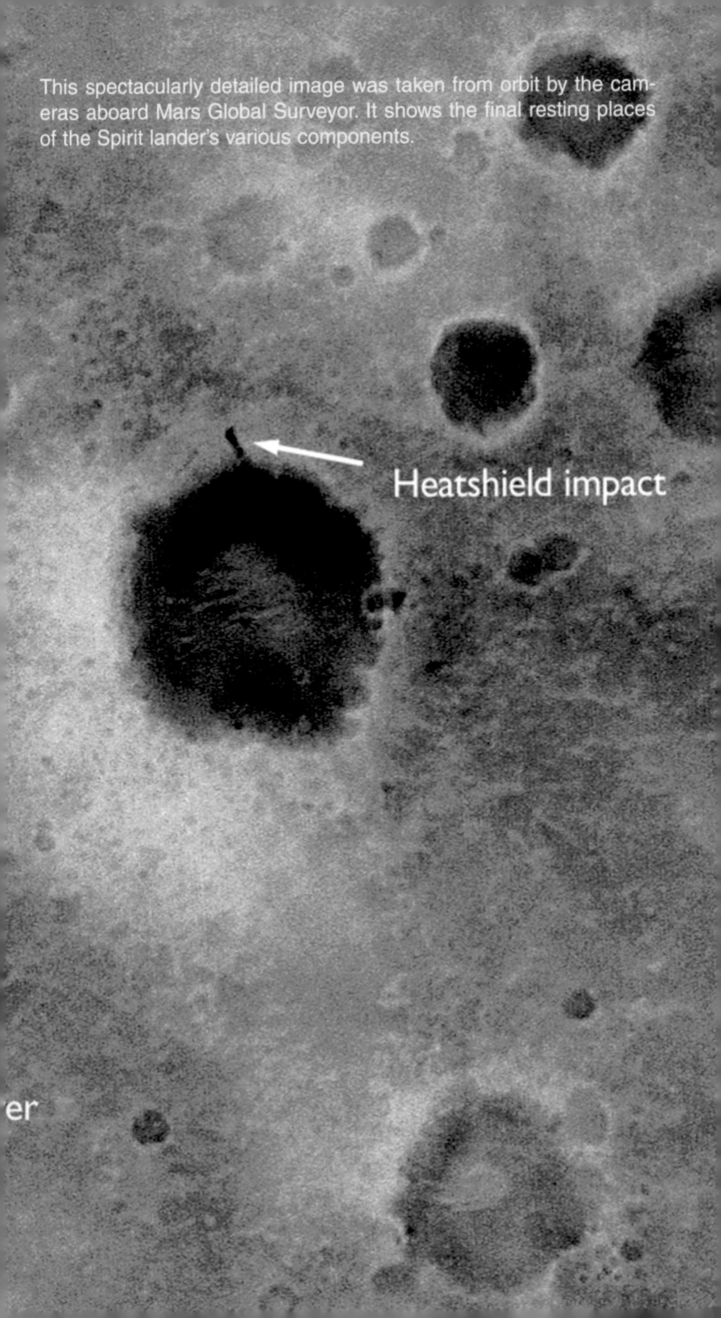

This spectacularly detailed image was taken from orbit by the cameras aboard Mars Global Surveyor. It shows the final resting places of the Spirit lander's various components.

Heatshield impact

er

Eagle Crater

Lander

Rover

Landing rock
blast effects

Backshell and
Parachute

Looking like a scene from a George Lucas movie NASA's Mars Exploration Rover Opportunity gained this view of its own heat shield during the rover's 325th martian day (Dec. 22, 2004). The main structure from the successfully used shield is to the far left. Additional fragments of the heat shield lie in the upper center of the image. The impact scar can be seen at right.

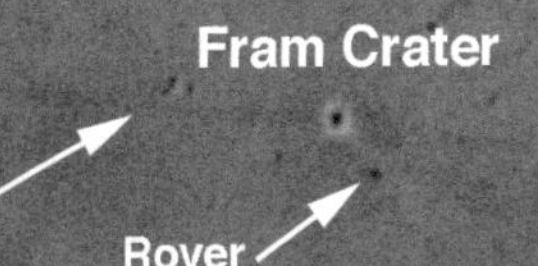

The image was acquired on April 26, 2004. The rover itself can be seen in this image -- an amazing accomplishment, considering that the orbiter was nearly 400 kilometers (nearly 250 miles) away at the time! Also visible and labeled on this image are the spacecraft's lander, backshell, parachute and heat shield, plus effects of its landing rockets. Incredibly the lander came to rest inside a small crater in what was referred to as a “hole-in-one”.

This high-resolution image captured by the Mars Exploration Rove
Opportunity's panoramic camera highlights the puzzling rock outcrop near th
rover's landing site. Opportunity investigated the outcrop with the suite of sc
entific instruments located on its robotic arm. These layered rocks measur

The rim and interior of a crater nicknamed "Bonneville" dominate this false
color mosaic of images taken by the panoramic camera of NASA's Mar
Exploration Rover Spirit. Spirit recorded this view onMarch 12, 2004.

On the 421st martian day of its time on Mars (March 31,2005), Rover Opportunity drove to within about 10 meters (33 feet) of a small crater called

only 10 centimeters (4 inches) tall, about the height of a street curb. Data from the panoramic camera's near-infrared, blue and green filters were combined to create this approximate, true-color image.

"Viking." the rover used its navigation camera to take images combined into this view of its new surroundings, including the crater. (false color)

The view is from Spirit's position known informally as "Larry's Lookout" alon the drive up "Husband Hill." The summit of Husband Hill is the far peak nea the center of this panorama and is about 200 meters (656 feet) away from th rover and about 45 meters (148 feet) higher in elevation.

Opportunity found an iron meteorite on Mars, the first meteorite of any type ever identified on another planet. The pitted, basketball-size object is mostly made of iron and nickel.

This view of the region near "Husband Hill" combines 243 images taken b Spirit over several martian days. It is an approximately true-color rendering.

SA Administrator Sean O'Keefe (left) and the Mars rover team are jubilant r the successful landing of the Opportunity rover. Left to right they are Dr. Weiler, Dr Charles Elachi, Pete Theisinger and Richard Cook

Another view of “Burns Cliff.”

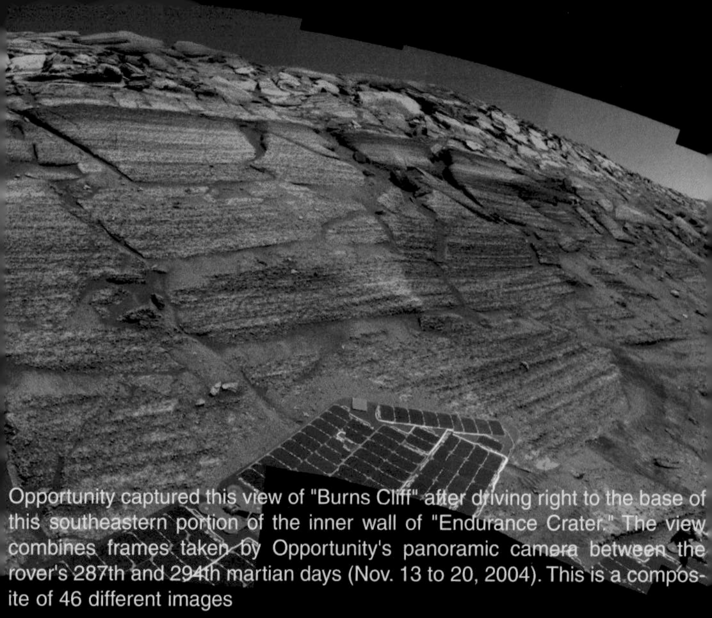

Opportunity captured this view of "Burns Cliff" after driving right to the base of this southeastern portion of the inner wall of "Endurance Crater." The view combines frames taken by Opportunity's panoramic camera between the rover's 287th and 294th martian days (Nov. 13 to 20, 2004). This is a composite of 46 different images

On the flank of "Husband Hill" inside Gusev Crater. This false-color view was assembled from frames taken by the Mars Exploration Rover Spirit's panoramic camera on the rover's 454th martian day

The Mars rover mission control team "storm" the press area at JPL after successfully landing the second rover on target. Vice-President Gore is at left.

The twin rovers Spirit and Opportunity performed way beyond their predicted "shelf-life" and discovered clear evidence that Mars was once wet, not in the form of Lowell's canals but perhaps in the form of a shallow salty sea.

Plans for a manned landing on Mars go back many decades. These two images come from a 1969 plan drawn up by Wernher von Braun.

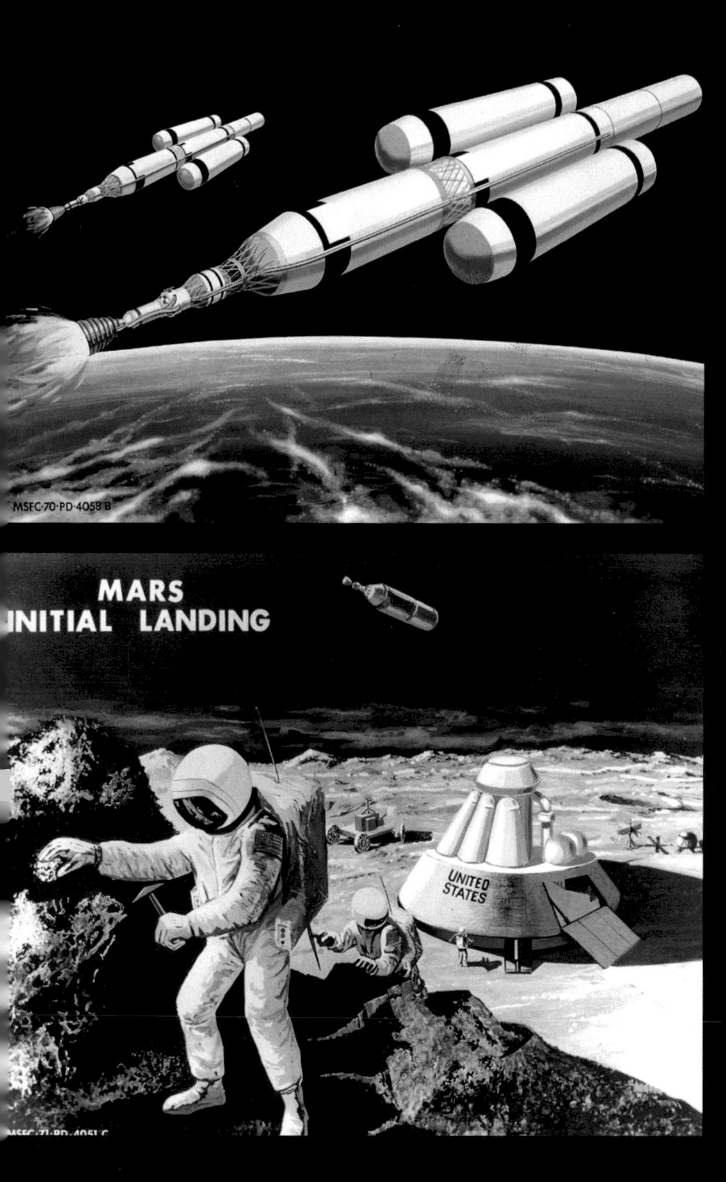

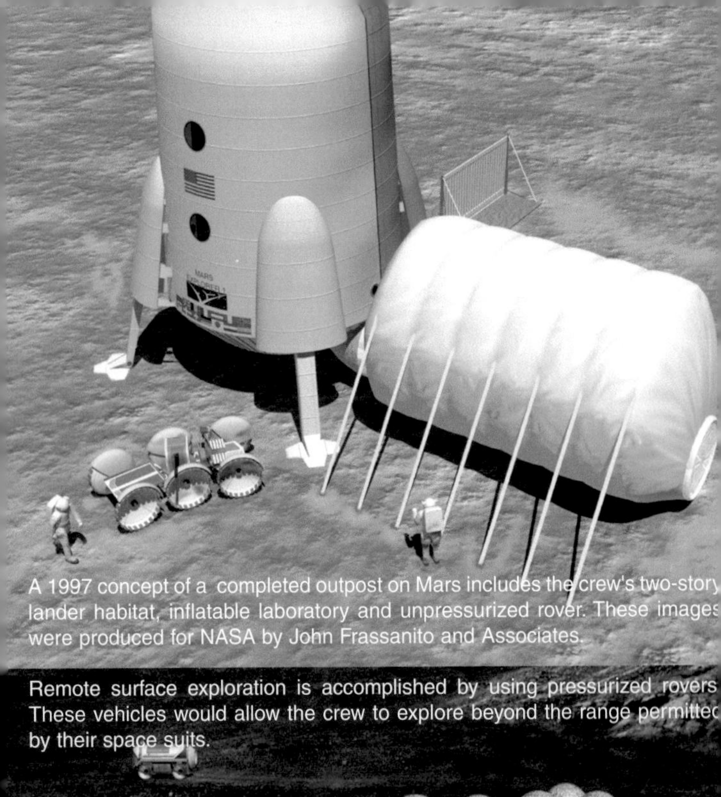

A 1997 concept of a completed outpost on Mars includes the crew's two-story lander habitat, inflatable laboratory and unpressurized rover. These images were produced for NASA by John Frassanito and Associates.

Remote surface exploration is accomplished by using pressurized rovers. These vehicles would allow the crew to explore beyond the range permitted by their space suits.

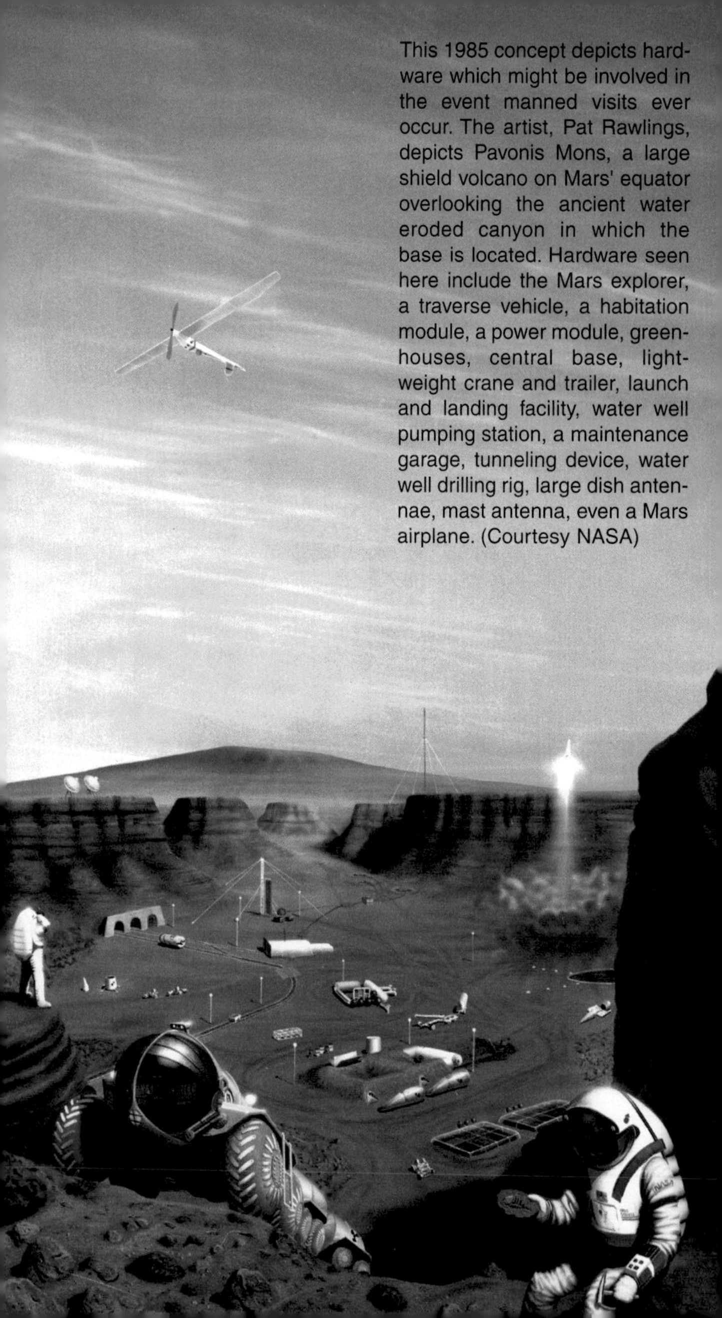

This 1985 concept depicts hardware which might be involved in the event manned visits ever occur. The artist, Pat Rawlings, depicts Pavonis Mons, a large shield volcano on Mars' equator overlooking the ancient water eroded canyon in which the base is located. Hardware seen here include the Mars explorer, a traverse vehicle, a habitation module, a power module, greenhouses, central base, lightweight crane and trailer, launch and landing facility, water well pumping station, a maintenance garage, tunneling device, water well drilling rig, large dish antennae, mast antenna, even a Mars airplane. (Courtesy NASA)

After landing on the Martia surface, the crew uses a unpressurized rover to unloa cargo and supplies neede for their stay on the red plan et. These images produce for NASA by John Frassani and Associates.

In front of a fully-fueled ascent vehicle waiting to return them to Earth, the Mars crew salutes all of the people and nations of the world that made the journey possible.

Sunset on Mars as seen by the Spirit rover.

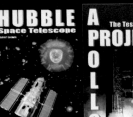

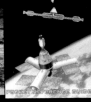